STUDENT SOLUTIONS MANUAL

Sarah Streett

SECOND EDITION

STATISTICS

THE ART AND SCIENCE OF LEARNING FROM DATA

Agresti • Franklin

Upper Saddle River, NJ 07458

Editorial Director: Christine Hoag
Editor-in-Chief, Mathematics & Statistics: Deirdre Lynch
Print Supplement Editor: Joanne Wendelken
Senior Managing Editor: Linda Mihatov Behrens
Associate Managing Editor: Bayani Mendoza de Leon
Project Manager: Barbara Mack
Art Director: Heather Scott
Supplement Cover Manager: Paul Gourhan
Supplement Cover Designer: Victoria Colotta
Operations Specialist: Ilene Kahn
Senior Operations Supervisor: Diane Peirano

© 2009 Pearson Education, Inc.
Pearson Prentice Hall
Pearson Education, Inc.
Upper Saddle River, NJ 07458

The author and publisher of this book have used their best efforts in preparing this book. These efforts include the development, research, and testing of the theories and programs to determine their effectiveness. The author and publisher make no warranty of any kind, expressed or implied, with regard to these programs or the documentation contained in this book. The author and publisher shall not be liable in any event for incidental or consequential damages in connection with, or arising out of, the furnishing, performance, or use of these programs.

Printed in the United States of America

10 9 8 7 6 5 4 3 2 1

ISBN-13: 978-0-13-603616-6

ISBN-10:   0-13-603616-3

Pearson Education Ltd., *London*
Pearson Education Australia Pty. Ltd., *Sydney*
Pearson Education Singapore, Pte. Ltd.
Pearson Education North Asia Ltd., *Hong Kong*
Pearson Education Canada, Inc., *Toronto*
Pearson Educación de Mexico, S.A. de C.V.
Pearson Education—Japan, *Tokyo*
Pearson Education Malaysia, Pte. Ltd.

Contents

1, STATISTICS: THE ART AND SCIENCE OF LEARNING FROM DATA   1

2. EXPLORING DATA WITH GRAPHS AND NUMERICAL SUMMARIES   4

3. ASSOCIATION: CONTINGENCY, CORRELATION, AND REGRESSION   21

4. GATHERING DATA   40
PART 1 REVIEW: GATHERING AND EXPLORING DATA   **49**

5. PROBABILITY IN OUR DAILY LIVES   53

6. PROBABILITY DISTRIBUTIONS   65

7. SAMPLING DISTRIBUTIONS   75
**PART 2 REVIEW: PROBABILITY, PROBABILITY DISTRIBUTIONS, AND
        SAMPLING DISTRIBUTIONS**   **82**

8. STATISTICAL INFERENCE: CONFIDENCE INTERVALS   84

9. STATISTICAL INFERENCE: SIGNIFICANCE TESTS ABOUT HYPOTHESES   94

10. COMPARING TWO GROUPS   106
PART 3 REVIEW: STATISTICAL INFERENCE   **121**

11. ANALYZING THE ASSOCIATION BETWEEN CATEGORICAL VARIABLES   127

12. ANALYZING THE ASSOCIATION BETWEEN QUANTITATIVE VARIABLES:
REGRESSION ANALYSIS   136

13. MULTIPLE REGRESSION   147

14. COMPARING GROUPS: ANALYSIS OF VARIANCE METHODS   160

15. NONPARAMETRIC STATISTICS   168
**PART 4 REVIEW: ANALYZING ASSOCIATIONS AND
        EXTENDED STATISTICAL METHODS**   **173**

Note:

Problems marked with the ⌨ symbol were solved using one of the text's applets, data files and a computer software program, or information obtained from the GSS, as indicated in the textbook.

Problems marked with the ♦♦ symbol have been denoted in the text as being more challenging.

Chapter 1
Statistics: The Art and Science of Learning from Data

SECTION 1.1: PRACTICING THE BASICS

1.1. **Aspirin and heart attacks:**
   a) Aspects of the study that have to do with design include the sample, the randomization of the halves of the sample to the two groups (aspirin and placebo), and the plan to obtain percentages of each group that have heart attacks.
   b) Aspects having to do with description include the actual percentages of the people in the sample who have heart attacks (i.e., 0.9% for those taking aspirin and 1.7% for those taking placebo).
   c) Aspects that have to do with inference include the use of statistical methods to predict whether the percentages for *all* male physicians would be lower for those taking aspirin than for those taking placebo.

⌨1.3 GSS and heaven:
   Yes, definitely: 64.8%; Yes, probably: 20.9%; No, probably not: 8.6%; No, definitely not: 5.8%

⌨1.5 GSS for subject you pick:
   The results for this item will be different depending on the topic that you chose.

SECTION 1.2: PRACTICING THE BASICS

1.7 Number of good friends:
   a) The sample is the 819 respondents to the General Social Survey question, "About how many good friends do you have?".
   b) The population is the American adult public.
   c) The statistic reported is the percentage of respondents having only 1 good friend (i.e., 6%).

1.9 EPA:
   a) The subjects in this study are cars – specifically, new Honda Accords.
   b) The sample is the few new Honda Accords that are chosen for the study on pollution emission and gasoline mileage performance.
   c) The population is all new Honda Accords.

1.11 Graduating seniors' salaries:
   a) These are descriptive statistics. They are summarizing data from a population – all graduating seniors at a given school.
   b) These analyses summarize data on a population – all graduating seniors at a given school; thus, the numerical summaries are best characterized as parameters.

1.13 Age pyramids as descriptive statistics:
   a) The graph shows fewer men and women as age increases. The bars on these graphs indicate thousands of people of a given gender and in a given age range. The very short bars toward the top indicate that there are very few men and women in their 70's and 80's in 1750.
   b) For every age range, the bars are much longer for both men and women in 2000 than in 1750.
   c) The bars for women in their 70's and 80's in 2000 are longer than those for men of the same age in the same year.
   d) The bars of people who were born right after World War II, now middle-aged, are the longest bars for both women and men.

1.15   National service:
   a) The populations are the same for the two studies. Two separate samples are taken from the same population.
   b) The sample proportions are not necessarily the same because the two random samples may differ by chance.

Section 1.3: Practicing the Basics

1.17   Data file for friends:
   The results for this exercise will be different for each person who does it. The data files, however, should all look like this:

Friend	Characteristic 1	Characteristic 2
1		
2		
3		
4		

   For each friend, you'll have a number or label under characteristics 1 and 2. For example, if you asked each friend for gender and hours of exercise per week, the first friend might have m (for male) under Characteristic 1, and 6 (for hours exercised per week) under Characteristic 2.

1.19   Internet poll:
   An Internet poll is not a random sample because every person in the population does not have the same chance of being in the sample. Some people do not have computers, others don't have Internet access, still others do not visit the site on which the poll is posted, and some choose not to participate. Those with computers and Internet access who frequently surf the web would have a much higher chance of being in this study than those who don't meet those criteria.

⌨1.21   Use a data file with software:
   See solution for 1.20 for format of data in MINITAB.

⌨1.23   Is a sample unusual?:
   It would be surprising to get a percentage that's more than 20 points from the true population percentage with a sample of 50 people. If you use the applet to conduct a simulation, you'll see that most of the time, the samples fall within 14 points of the true population percentage – from about 56 to 84.

Chapter Problems: Practicing the Basics

1.25   ESP:
   a) The population of interest is all American adults (the population from which this sample was taken).
   b) The sample data are summarized by giving a proportion of all subjects (0.638) who said that they had at least one such experience, rather than giving the individual data points for all 3887 sampled subjects.
   c) We might want to make an inference about the population with respect to the proportion who had had at least one ESP experience. We would use the sample proportion to estimate the population proportion.

1.27   Bush vs. Kerry in other countries:
   a) The results summarize sample data because not everyone in each country was polled.
   b) The percentages reported here are descriptive in that they describe the exact percentages of the samples polled who preferred Kerry or Bush.
   c) The inferential aspect of this analysis is that the BBC report is implying that these percentages provide information about the general population of each of these countries. The margin of error for the sample percentage gives information about the likely range in which the percentages fall in each of these countries.

1.29 **Marketing study**:
a) For the study on the marketing of CD's, the population is all customers to whom catalogs could be sent, and the sample is the 500 customers to whom catalogs actually are sent.
b) Example 4 suggests that we might determine that the average sales per person equaled $4.
This would be a descriptive statistic in that it describes the average sales per person in the sample of 500 customers. If one were to use this information to make a prediction about the population, this would be an inferential statistic.

1.31 **Use of inferential statistics?**:
c) to make predictions about populations using sample data

CHAPTER PROBLEMS: CONCEPTS AND INVESTIGATIONS

1.33 **Statistics in the news**:
If your article has numbers that summarize for a given group (sample or population), it's using descriptive statistics. If it uses numbers from a sample to predict something about a population, it's using inferential statistics.

1.35 **Surprising ESP data?**:
This result would be very surprising with such a large sample. You'll notice that when you use the applet to simulate this study, you will get a sample proportion as large as 0.638, when the true proportion is 0.20, only VERY rarely. With such a large sample, if randomly selected, you'd expect a sample proportion very close to the population proportion.

Chapter 2
Exploring Data with Graphs and Numerical Summaries

SECTION 2.1: PRACTICING THE BASICS

2.1 **Categorical/quantitative difference:**
   a) Categorical variables are those in which observations belong to one of a set of categories, whereas quantitative variables are those on which observations are numerical.
   b) An example of a categorical variable is religion. An example of a quantitative variable is temperature.

2.3 **Identify the variable type:**
   a) quantitative
   b) categorical
   c) categorical
   d) quantitative

2.5 **Discrete/continuous:**
   a) A discrete variable is a quantitative variable for which the possible values are separate values such as 0, 1, 2, …. A continuous variable is a quantitative variable for which the possible values form an interval.
   b) Example of a discrete variable: the number of children in a family (a given family can't have 2.43 children).
   Example of a continuous variable: temperature (we <u>can</u> have a temperature of 48.659).

2.7 **Discrete or continuous 2:**
   a) continuous
   b) discrete
   c) discrete
   d) continuous

SECTION 2.2: PRACTICING THE BASICS

2.9 **Environmental protection:**
   a)

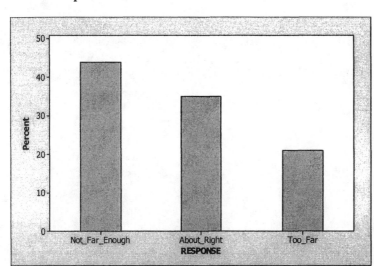

   b) It's much easier to sketch bar charts relatively accurately.
   c) The advantage of using one of these visual displays to summarize the results is that we can get a better sense of the data when we can see the sizes of the various categories as opposed to just reading the numbers.

2.11   Weather stations:
a) The slices of the pie portray categories of a variable (i.e., regions).
b) The first number is the frequency, the number of weather stations in a given region. The second number is the percentage of all weather stations that are in this region.
c) It is easier to identify the mode using a bar graph than using a pie chart because we can more easily compare the heights of bars than the slices of a piece of pie. For example, in this case, the slices for Midwest and West look very similar in size, but it would be clear from a bar graph that West was taller in height than Midwest.

2.13   Shark attacks worldwide:
The following charts use proportions.
(i) Alphabetically:

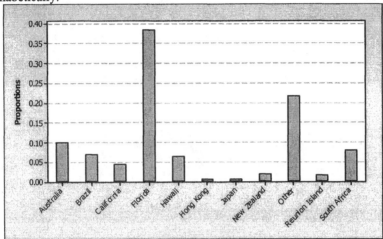

(ii) Pareto chart:

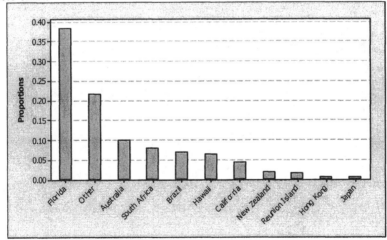

The Pareto chart is more useful than the chart arranged alphabetically because we can easily compare regions and see what outcomes occurred most frequently.

2.15   **eBay prices:**
a)   17 | 8
        18 |
        19 | 9
        20 | 0
        21 | 00
        22 | 55558
        23 | 25
        24 | 00056669
        25 | 00055

b)   This plot gives us an overview of all the data. We see clearly that the prices fall between $178 and $255, with more PDAs selling at the upper end of this range.

2.17   **Fertility rates:**
a)   1 | 3333445677778899
     2 | 04
     A disadvantage of this plot is that it is too compact making it difficult to visualize where the data fall.

b)   1 | 333344
     1 | 5677778899
     2 | 04

c)

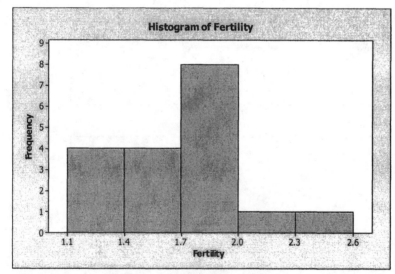

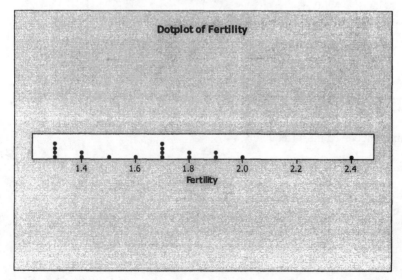

2.19   **Leaf unit**:
   a)   The observation in the first row of the plot is 1000 milligrams.
   b)   The sugar outcome that occurs most frequently is 3000 milligrams.

2.21   **Histogram for sugar**:
   a)   The intervals are 1 to 3, 3 to 5, 5 to 7, 7 to 9, 9 to 11, 11 to 13, 13 to 15, and 15 to 17.
   b)   The distribution is bimodal. This is likely due to the fact that grams of sugar for both adult and child cereals are included. Child cereals, on average, have more sugar than adult cereals have.
   c)   The dot and stem-and-leaf plots allow us to see all the individual data points; we cannot see the individual value sin the histogram.
   d)   The label of the vertical axis would change from Frequency to Percentage, but the relative differences among bars would remain the same. For example, the frequency for the lowest interval is 2; that would convert to 10% of all data points. The frequency for the second interval is 4, which would convert to 20%. In both cases, the second interval is twice as big as the first.

2.23   **Shape of the histogram**:
   a)   Assessed value houses in a large city – skewed to the right (a long right tail) because of some very expensive homes
   b)   Number of times checking account overdrawn in the past year for the faculty at the local university – skewed to the right because of the few faculty who overdraw frequently
   c)   IQ for the general population – symmetric because most would be in the middle, with some higher and some lower; there's no reason to expect more to be higher or lower (particularly because IQ is constructed as a comparison to the general population's "norms")
   d)   The height of female college students – symmetric because most would fall in the middle, going down to a few short students and up to a few tall students

2.25   **How often do students read the newspaper?**
   a)   This is a discrete variable because the value for each person would be a whole number. One could not read a newspaper 5.76 times per week, for example.
   b)   (i) The minimum response is zero.
        (ii) The maximum response is nine.
        (iii) Two students did not read the newspaper at all.
        (iv) The mode is three.
   c)   This distribution is unimodal and somewhat skewed to the right.

2.27   **Central Park temperatures:**
   a)   The distribution is somewhat skewed to the left; that is, it has a tail to the left.
   b)   A time plot connects the data points over time in order to show time trends – typically increases or decreases over time. We cannot see these changes over time in a histogram.
   c)   A histogram shows us the number of observations at each level; it is more difficult to see how many years had a given average temperature in a time plot. We also can see the shape of the distribution of temperatures from the histogram but not from the time plot.

⌨2.29 Warming in Newnan, GA?:
Overall, the time plot (below) does seem to show a decrease in temperature over time.

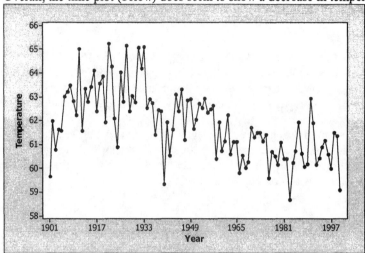

SECTION 2.3: PRACTICING THE BASICS

2.31 More on CO_2 emissions:

a) Mean: $\bar{x} = \dfrac{\sum x}{n} = (1892+1356+520+506+364+265+230)/7 = 733.3$

Median: Find the middle value [(n+1)/2=(7+1)/2=4]
    230 265 364 506 520 1356 1892
    The median is 506.

b) Some of these countries, including the U.S. and China, have much larger populations than others. One would expect total CO_2 emissions to increase as population size increased. It would be more useful to know the level of CO_2 emissions per person.

2.33 Income and race:
For both blacks and whites, the income distributions are skewed to the right. When this happens, the means are higher than are the medians, because they are affected by the size of the outliers (a few very rich individuals).

2.35 Cereal sodium:
The sodium value of zero is an outlier. It causes the mean to be skewed to the left, but its actual value does not affect the median.

2.37 Public transportation – center:
a) The mean is 2, the median is 0, and the mode is zero. Thus, the average score is 2, the middle score is zero (indicating that the mean is skewed by outliers), and the most common score also is zero.

Mean: $\bar{x} = \dfrac{\sum x}{n} = (0+ 0+4+ 0+ 0+0+10+0+6+0)/10 = 2$

Median: middle score of 0, 0, 0, 0, 0, 0, 0, 4, 6, and 10
Mode: the most common score is zero.

b) Now the mean is 10, but the median is still 0.

Mean: $\bar{x} = \dfrac{\sum x}{n} = (0+0+4+0+0+0+10+0+6+0+90)/11 = 2$

Median: middle score of 0, 0, 0, 0, 0, 0, 0, 4, 6, 10, and 90
The median is not affected by the magnitude of the highest score, the outlier. Because there are so many zeros, even though we've added one score, the median remains zero. The mean, however, is affected by the magnitude of this new score, an extreme outlier.

2.39   Student hospital costs:
a) The vast majority of students did not have any hospital stays last year. Thus, the most common score (mode), and the middle score (median), will be zero. There will be several students with hospital stays, however, and there likely will be a few with very expensive hospital stays. The money spent by these students will make the mean positive.
b) Another variable that would have this property is the number of aquarium fish in one's home. Most people don't have any, but there are a few people with fish-filled tanks. Another example is the number of times arrested in the previous year.

2.41   Canadian income:
This distribution is skewed to the right. The fact that the mean is higher than the median indicates that there are extremely high incomes that are affecting the mean, but not the median.

2.43   European fertility:
a) The median fertility rate is 1.7. Thus, about half of the countries listed have mean fertility rates at or below 1.7 with the remaining countries having fertility rates above 1.7.
b) The mean of the fertility rates is 1.65.
c) Since the population of adult women can vary greatly among the countries, it is necessary to calculate an overall fertility rate for the country in order to make comparisons. This rate is found by calculating the mean number of children per adult woman. The mean for a variable need not be one of the possible values for the variable. Although the number of children born to each adult woman is a whole number, the mean number of children born per adult woman need not be a whole number. For example, the mean number of children per adult woman is considerably higher in Mexico than in Canada.

2.45   Knowing homicide victims:
a) The mean is 0.17.

$\sum x/n = (3944(0) + 279(1) + 97(2)+40(3)+30(4.5))/4390 = 0.17$.

b) The median is the average of the two middle scores. With 4390 scores, the median falls between scores 2195 and 2196. Thus, the median is 0.
c) The median would still be 0, because there are still 2200 people who gave 0 as a response. The mean would now be 1.95.

$\sum x/n = (2200(0) + 279(1) + 97(2)+40(3)+1804(4.5))/4390 = 1.95$.

d) The median is the same for both because the median ignores much of the data. The data are highly discrete; hence, a high proportion of the data falls at only one or two values. The mean is better in this case because it uses the numerical values of all of the observations, not just the ordering.

SECTION 2.4: PRACTICING THE BASICS

2.47   Sick leave:
   a)   The range is six; this is the distance from the smallest to the largest observation. In this case, there are six days separating the fewest and most sick days taken (6-0=6).
   b)   The standard deviation is the typical distance of an observation from the mean (which is 1.25).

$$s^2 = \frac{\sum(x-\overline{x})^2}{n-1} = ((0\text{-}1.25)^2 + (0\text{-}1.25)^2 + (0\text{-}1.25)^2 + (0\text{-}1.25)^2 + (0\text{-}1.25)^2 + (0\text{-}1.25)^2 + (4\text{-}1.25)^2 + (6\text{-}1.25)^2)/7 = 39.5/7 = 5.64$$

$$s = \sqrt{s^2} = \sqrt{5.643} = 2.38$$

The standard deviation of 2.38 indicates a typical number of sick days taken is 2.38 days from the mean of 1.25.
   c)   Redo (a) and (b)
      a.   The range is sixty; this is the distance from the smallest to the largest observation. In this case, there are sixty days separating the fewest and most sick days taken (60-0=60).
      b.   The standard deviation is the typical distance of an observation from the mean (which is 8).

$$s^2 = \frac{\sum(x-\overline{x})^2}{n-1} = ((0\text{-}8)^2 + (0\text{-}8)^2 + (0\text{-}8)^2 + (0\text{-}8)^2 + (0\text{-}8)^2 + (0\text{-}8)^2 + (4\text{-}8)^2 + (60\text{-}8)^2)/7 = 3104/7 = 443.43$$

$$s = \sqrt{s^2} = \sqrt{443.43} = 21.06$$

The standard deviation of 21.06 indicates a typical number of sick days taken is 21.06 days from the mean of 8.
The range and mean both increase when an outlier is added.

2.49   Life expectancy including Russia:
We would expect the standard deviation to be larger since the value for Russia is significantly smaller than the rest of the group adding additional spread to the data. The standard deviation including Russia is, in fact, 3.56.

2.51   Exam standard deviation:
The most realistic value is 12. There are problems with all the others.
-10: We can't have a negative standard deviation.
0:   We know that there is spread because the scores ranged from 35 to 98, so the standard deviation is not 0.
3:   This standard deviation seems very small for this range.
63: This standard deviation is too large for a typical deviation. In fact, <u>no</u> score differed from the mean by this much.

2.53   Histograms and standard deviation:
   a)   The sample on the right has the largest standard deviation, and the sample in the middle has the smallest.
   b)   The Empirical Rule is relevant only for the distribution on the left because the distribution is bell-shaped.

2.55   Female body weight:
   a)   95% of weights would fall within two standard deviations from the mean – between 99 and 167.
   b)   An athlete who is three standard deviations above the mean would weight 184 pounds. This would be an unusual observation because typically all or nearly all observations fall within three standard deviations from the mean. In a bell-shaped distribution, this would likely be about the highest score one would obtain.

2.57 **Empirical rule and skewed, highly discrete distribution**:
a) One standard deviation from the mean: between -0.21 and 0.59
Two standard deviations from the mean: between -0.61 and 0.99
Three standard deviations from the mean: between -1.01 and 1.39

	Observations	Predicted by Empirical Rule
One standard deviation from the mean	81.7%	68%
Two standard deviations from the mean	81.7%	95%
Three standard deviations from the mean	99.5%	About 100%

There are more observations within one standard deviation of the mean and fewer within two standard deviations than would be predicted by the Empirical Rule.

b) The Empirical Rule is only valid when used with data from a bell-shaped distribution. This is not a bell-shaped distribution; rather, it is highly skewed to the right. Most observations have a value of 0, and hardly any have the highest value of 2.

2.59 **How many friends?**:
a) The standard deviation is larger than the mean; in addition, the mean is higher than the median. In fact, the lowest possible value of 0 is only 7.4/11.0 = 0.67 standard deviations below the mean. These situations occur when the mean and standard deviation are affected by an outlier or outliers. It appears that this distribution is skewed to the right.

b) The Empirical Rules does not apply to these data because they do not appear to be bell-shaped.

2.61 **EU data file**:
a)

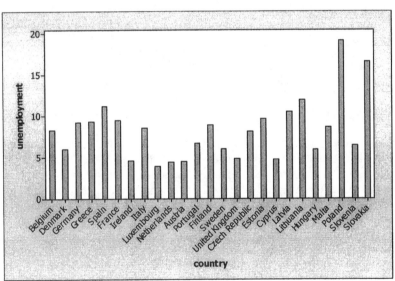

b) Variable   StDev
   unemployment 3.695
   This is a typical deviation from the mean unemployment for EU nations.

SECTION 2.5: PRACTICING THE BASICS

2.63   Vacation days:
a)   Median: Find the middle value of 13, 25, 26, 28, 34, 35, 37, 42
The median is 31, the average of the two middle values, 28 and 34.
b)   The first quartile is the median of 13, 25, 26, and 28. The first quartile is 25.5, the average of the two middle values, 25 and 26.
c)   The third quartile is the median of 34, 35, 37, and 42. The third quartile is 36, the average of the two middle values.
d)   25% of countries have residents who take fewer than 25.5 vacation days, half of countries have residents who take fewer than 31 vacation days, and 75% of countries have residents who take fewer than 36 vacation days per year. The middle 50% of countries have residents who take an average of between 25.5 and 36 vacation days annually.

2.65   Female strength:
a)   One fourth had maximum bench press less than 70 pounds, and one fourth had maximum bench press greater than 90 pounds.
b)   The mean and median are the about the same, and the first and third quartiles are equidistant from the median. These are both indicators of a roughly symmetric distribution.

2.67   Ways to measure spread:
a)   The range is even more affected by an outlier than is the standard deviation. The standard deviation takes into account the values of all observations and not just the most extreme two.
b)   With a very extreme outlier, the standard deviation will be affected both because the mean will be affected and because the deviation of the outlier (and its square) will be very large. The IQR would not be affected by such an outlier.
c)   The standard deviation takes into account the values of all observations and not just the two marking 25% and 75% of observations.

2.69   Sick leave:
a)   The range is six; this is the distance from the smallest to the largest observation. In this case, there are six days separating the fewest and most sick days taken (6-0=6).
b)   The interquartile range is the difference between Q3 and Q1. IQR=Q3-Q1=2.
c)   Redo (a) and (b).
   a.   The range is sixty; this is the distance from the smallest to the largest observation. In this case, there are sixty days separating the fewest and most sick days taken (60-0=60).
   b.   Q1, the median of all scores below the median, is still 0. Q3, the median of all scores above the median, is still 2 (the average of 0 and 4). The interquartile range remains the same: IQR=Q3-Q1=2. The IQR is least affected by the outlier because it doesn't take the magnitudes of the two extreme scores into account at all, whereas the range and *s* do.

2.71   Infant mortality Europe:
Q1 is the median of the lower half of the sorted data: 3, 3, 3, 4, 4, 4 and 4. It is 4. Q2, the median, is the average of the middle two data values (4+4)/2=4. Q3 is the median of the upper half of the sorted data: 4, 4, 4, 4, 5, 5 and 5. It is 4.

2.73   Central Park temperature distribution revisited:
a)   We would expect it to be skewed to the left because the maximum is closer to the median than is the minimum.
b)   Numbers are approximate:
Minimum: 50;   Q1: 53;   Median: 54.2;   Q3: 55;   Maximum: 57.3
These approximations support the premise that the distribution is skewed to the left if it is skewed. The median is closer to the maximum and Q3 than it is to the minimum and Q1.

2.75 **Public transportation:**

a) Minimum: 0
   Q1: 0
   Median: 0
   Q3: 4
   Maximum: 10

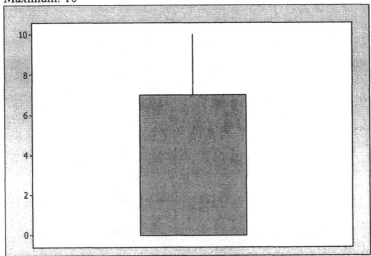

b) Q1 and the median share the same line in the box because so many employees have a score of zero that the middle score of the whole set of data is zero and the middle score of the lower half of the data also is zero.

c) There is no whisker because the minimum score also is zero. This situation resulted because there are so many people with the lowest score.

2.77 **European Union unemployment rates:**

a) In a box plot, Q1, 4.5, and Q3, 7.8, would be at the outer edges of the box. The whiskers would extend on the left to the minimum, 3.2, and on the right to the maximum, 8.7.

b) The score is 1.33 standard deviations above the mean, and thus, is not an outlier according to the three standard deviation criterion.

$$z = \frac{x - \bar{x}}{s}, z = \frac{8.7 - 6.3}{1.8} = 1.33.$$

c) A z-score of 0 indicates that the country's unemployment rate is zero standard deviations from the mean; hence, the unemployment rate is equal to the mean. In this case, a country with an unemployment rate of 6.3 would have a z-score of 0.

2.79 **Female heights:**

a) $z = \frac{x - \bar{x}}{s}, z = \frac{56 - 65.3}{3.0} = -3.1.$

b) The negative sign indicates that the height of 56 inches is below the mean.

c) Because the height of 56 inches is more than three standard deviations from the mean, it is a potential outlier.

2.81 Florida students again:
  a)   The distribution depicted in the boxplot below is skewed to the right. The right whisker is longer, and there are a few outliers (shown by asterisks) representing large values.

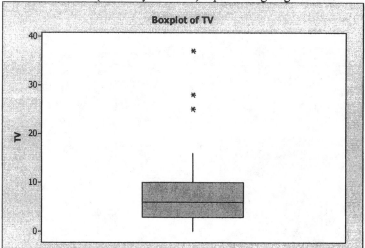

  b)   The 1.5(IQR) criterion would indicate that all data should fall between about -7.5 and 20.5 (based on IQR = 10 – 3 = 7; 1.5(IQR)1.5 = 10.5). Because some data points fall beyond this range, it appears that there are potential outliers.

2.83 CO2 comparison:
  a)   The outlier shown in the box plot for Europe is around 22.
  b)   Based on the box plot, we would predict the shape of the distribution of carbon dioxide emissions for South America to be skewed to the right since the distance between Q3 and Q2 is much larger than between Q1 and Q2 and the upper whisker is also much longer than the lower whisker.
  c)   Carbon dioxide emissions tend to be much higher in Europe than South America. The smallest level reported in Europe is close to the third quantile of the levels reported in South America. Thus, roughly 75% of the levels reported in South America are less than the smallest level reported in Europe.

SECTION 2.6: PRACTICING THE BASICS

2.85 Market share for food sales:
  a)   One problem with this chart is that the percentages do not add up to 100. Second, the Tesco slice seems too large for 27.2%. A third problem is that contiguous colors are very similar, particularly the two white sections. This increases the difficulty in easily reading this chart.
  b)   It would be easier to identify the mode with a bar graph because one would merely have to identify the highest bar.

2.87 Terrorism and war in Iraq:
  a)   This graph is misleading. Because the vertical axis does not start at 0, it appears that six times as many people are in the "no, not" column than in the "yes, safer" column, when really it's not even twice as many.
  b)   With a pie chart, the area of each slice represents the percentage who fall in that category. Therefore the relative sizes of the slices will always represent the relative percentages in each category.

2.89 Spending on drugs:
The slices do not seem to have the correct sizes, for instance the slice with 6.9% seeming larger than the slice with 7.2%.

CHAPTER PROBLEMS: PRACTICING THE BASICS

2.91 **Categorical or quantitative?**:
a) Number of children in family: quantitative
b) Amount of time in football game before first points scored: quantitative
c) Choice of major (English, history, chemistry, ...): categorical
d) Preference for type of music (rock, jazz, classical, folk, other): categorical

2.93 **Immigration into U.S.**:
a) "Place of birth" is categorical.
   Percentages:

Europe	13.7%
Caribbean	9.6%
Central America	37.6%
South America	6.1%
Asia	25.4%
Other	7.6%
Total	100%

b) These categories are not ordered. Thus, the mode would be the most sensible measure. In this case, the mode is Central America.
c) In a Pareto chart, the place of birth categories would be organized from the one with the highest percentage (i.e., Central America) to the one with the lowest percentage (i.e., South America). The advantage of a Pareto chart is that it's much easier to make comparisons and to identify easily the most common outcomes (i.e., to identify which places of birth have higher percentages of current foreign-born Americans).

2.95 **Chad voting problems**:
a) We first locate the dot directly above 11.6% on the horizontal x axis. We then look at the vertical y axis across from this point to determine the label for that dot: optical, two column. This tells us that the over-vote was highest among those using optical scanning with a two-column ballot.
b) We first locate the dots above the lowest percentages on the x axis. We then determine the labels across from these dots on the y axis to determine the lowest two combinations: optical, one column, and votomatic, one column. Thus, the lowest over-voting occurred when voters had a ballot with only one column that was registered either using optical scanning or votomatic (manual punching of chads).
c) We could summarize these data further by using a bar for each combination: optical, one column; optical, two column; votomatic, one column, etc. For each bar, we could then plot the average over-vote of all counties in that category. To do this, we would need the exact percentages of each county in each category.

2.97 **Newspaper reading**:
a)

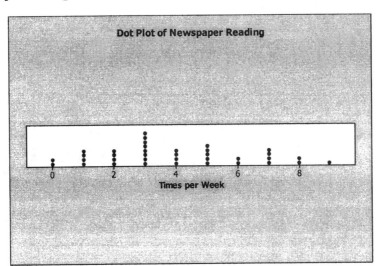

   b)   Stem-and-leaf of Times per Week N = 36
        Leaf Unit = 0.10
         0 00
         1 0000
         2 0000
         3 00000000
         4 0000
         5 00000
         6 00
         7 0000
         8 00
         9 0
        The leaf unit is identified above. The stems are the whole numbers, 0 through 9.
   c)   The median is the middle number. There are 36 numbers, so the median is between 18 and 19 which have
        the values 3 and 4, respectively. Thus, the median is 3.5.
   d)   The distribution is slightly skewed to the right.

2.99   **Enchiladas and sodium:**
     a)

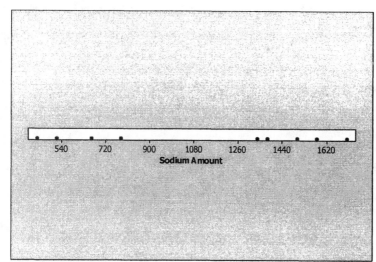

     b)   A dot plot allows one to see the individual scores.
     c)   Stem-and-leaf of Sodium Amount N = 9
          Leaf Unit = 10
           4   4
           5   2
           6   6
           7   8
           8
           9
          10
          11
          12
          13 37
          14
          15 07
          16
          17 0
     d)   This is a bimodal distribution. There are scores on either end of the distribution, but none in the middle.

2.101 **What shape do you expect?:**
   a) Number of times arrested in past year – skewed to the right because most values are at 0 but there are some large values.
   b) Time needed to complete difficult exam (maximum time is 1 hour) – skewed to the left because most values are at 1 hour or slightly less, but some could be quite a bit less.
   c) Assessed value of home – skewed to the right because there are some extremely large values.
   d) Age at death – skewed to the left because most values are high, but some very young people die.

2.103 **Median vs. mean income**:
   We would expect the mean to have been larger because of right skew – the very few Bill Gates and Oprah Winfreys of the world. The magnitude of their income would affect the mean, but not the median.

2.105 **Baseball salaries**:
   $1.2 million was the mean, and $500,000 was the median because the mean was affected by outliers and severe right skew. There are a few valuable players who receive exorbitant salaries (e.g., A-Rod), whereas the typical player is paid much less (although still a lot by most people's standards!). The exorbitant salaries of the few affect the mean, but not the median.

2.107 **Lengths of hikes:**
   a) One example is 1, 2, 4, 6, 7. Both the mean and median are 4.
   b) One example is 2, 5, 2, 6 and 3.

2.109 **What does *s* equal?:**
   a) Given the mean and range, the most realistic value is 12. -10 is not realistic because standard deviation must be 0 or positive. Given that there is a large range, it is not realistic that there would be almost no spread; hence, the standard deviation of 1 is unrealistic. 60 is unrealistically large; the whole range is hardly any more than 60.
   b) -20 is impossible because standard deviations must be nonnegative.

2.111 **Energy and water consumption:**
   a) The distribution is likely skewed to the right because the maximum is much farther from the mean than the minimum is, and also because the standard lowest possible value of 0 is only 780/506 = 1.54 standard deviations below the mean.
   b) The distribution is likely skewed to the right because the standard deviation is almost as large as the mean, and the smallest possible value is zero, only 1.15 standard deviation below the mean.

2.113 **Student heights:**
   a) For a bell-shaped distribution, such as the heights of all men, the Empirical Rule states that all or nearly all scores will fall within three standard deviations of the mean ($\bar{x} \pm 3s$). In this case, that means that nearly all scores would fall between 62.2 and 79.6. In this case, almost all men's scores do fall between these values.
   b) The center for women is about five inches less than the center for men. The spread, however, is very similar. These distributions are likely very similar in shape; they are just centered around different values.
   c) The lowest score for men is 62. This would have a *z*-score of $z = \dfrac{x - \bar{x}}{s}$ =(62-70.9)/2.9= -3.07. Thus, it falls 3.07 standard deviations below the mean.

2.115 Cost of books:

a) The sorted prices for the hardcover fiction books are: 19.95, 22, 24, 25, 25, 25.95, 26, 26, 26.95, 27, 27.95, 28.95, 29.95, 30 and 35. The minimum price is $19.95. The median, Q2, is the middle value, $26. Q1 is the median of the lower half of the prices, $25. Q3 is the median of the upper half of the prices, $28.95. The maximum price is $35.

b) The sorted prices for the paperback fiction books are: 7.99, 9, 12.95, 12.99, 13.95, 13.95, 13.95, 14, 14, 14.95, 14.95, 14.95, 14.95, 15 and 15. The minimum price is $7.99. The median, Q2, is the middle price, $14. Q1 is the median of the lower half of the sorted prices, $12.99. Q3 is the median of the upper half of the sorted prices, $14.95. The maximum price is $15.

2.117 Cereal sugar values:

a) Numbers are approximate:
Minimum: 1; Q1: 3 ; Median: 10; Q3: 12; Maximum: 15

b) Because the median is much closer to Q3 and the maximum than it is to Q1 or the minimum, it appears that this distribution is skewed to the left.

c) $z = \dfrac{x - \overline{x}}{s} = (1-8.20)/4.56 = -1.58$. This sugar value falls 1.58 standard deviations below the mean.

2.119 Temperatures in Central Park

a) The distribution appears to be roughly symmetric, although perhaps slightly skewed to the left. The mean and median are almost the same, bolstering the contention that the data are normal. There is very little spread, as evidenced by the small standard deviation in comparison with the mean.

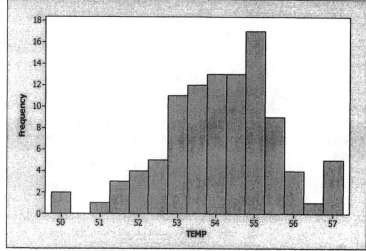

b) Outliers are indicated with asterisks outside of the whiskers. This box plot does not indicate any outliers.

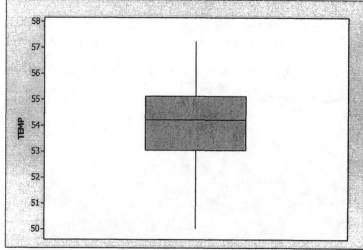

   c)   Minimum: 50.040 (rounds to 50.0)
       Q1: 53.055 (rounds to 53.1)
       Median: 54.215 (rounds to 54.2)
       Q3: 55.113 (rounds to 55.1)
       Maximum: 57.220 (rounds to 57.2)
       The median is fairly equidistant from Q1 and Q3, and also is fairly equidistant from the minimum and the maximum; this suggests a roughly symmetric distribution.

   d)   The mean is 54.097 (rounds to 54.1) and the standard deviation is 1.447 (rounds to 1.4). The mean is the average of all of the data points. The standard deviation is the amount that a typical score varies from the mean.

2.121   Health insurance:

   a)   The distribution is probably skewed to the right because the distance of Q3 from the median and from Q3 to the maximum is longer than the distance of Q1 from the median and from Q1 to the minimum.

   b)   The most plausible value for the standard deviation of this distribution is 4. The middle 50% of scores fall within a range of 5.4%., making it plausible that the typical score would deviate four percentage points from the mean. We cannot have a negative percentage point, so -16 is not plausible. We know that there is variation, so 0 is not plausible. The whole range is not much more than 15; thus, 15 and 25 are implausibly large for the standard deviation of this distribution.

2.123   High school graduation rates:

   a)   The range is the difference between the lowest and highest scores: 92.3-78.3=14.
       The interquartile range (IQR) is the difference between scores at the 25th and 75th percentiles: IQR=Q3-Q1=88.8-83.6=5.2.

   b)   1.5(IQR) = 7.8 from Q1 or Q3; this criterion suggests that potential outliers would be those scores less than 75.8 and greater than 96.6. There are no scores beyond these values, and so it would not indicate any potential outliers.

2.125   Blood pressure:

   a) $z = \dfrac{x - \bar{x}}{s} = (140\text{-}121)/16 = 1.19$

       A z-score of 1.19 indicates that a person with a blood pressure of 140, the cutoff for having high blood pressure, falls 1.19 standard deviations above the mean.

   b)   About 95% of all values in a bell-shaped distribution fall within two standard deviations of the mean – in this case, 32. Subtracting two times the standard deviation from the mean, and adding two times the standard deviation to the mean tells us that about 95% of systolic blood pressures fall between 89 and 153.

2.127   Who was Roger Maris:

   a) $z = \dfrac{x - \bar{x}}{s} = (5\text{-}22.92)/15.98 = -1.12$

       $z = \dfrac{x - \bar{x}}{s} = (61\text{-}22.92)/15.98 = 2.38$

       Neither the minimum nor the maximum score reaches the criterion for a potential outlier of being more than three standard deviations from the mean (i.e., having a z-score less than -3 or greater than 3). Thus, there are no potential outliers according to three standard deviation criterion.

   b)   The maximum is much farther from the mean and median than is the minimum, an indicator that the distribution might not be bell-shaped. Moreover, the lowest possible value of 0 is only 22.92/15.98 = 1.43 standard deviations below the mean.

   c)   Based on the criteria noted above, this is not unusual. It does not even come close to meeting the three standard deviation criterion for a potential outlier and therefore is not an unusual number of homeruns for Roger Maris.

       $z = \dfrac{x - \bar{x}}{s} = (13\text{-}22.92)/15.98 = -0.62$

CHAPTER PROBLEMS: CONCEPTS AND INVESTIGATIONS

⌨**2.129 How much spent on haircuts?:**
The responses will be different for each student depending on the methods used.

⌨**2.131 Google trend:**
The response to this exercise will be different for each student.

2.133   Back-to-back stem-and-leaf plot:
   a)   The two distributions are different. The median for the adult cereals is 4000 milligrams, with a shape skewed to the right. The median for the child cereals is between 1150 milligrams with a shape skewed to the left. In addition, the stems for the child cereals are more spread out than those of the adult cereals.
   b)   Adult   Child
                0| 0 |
                 | 0 |7
                4| 1 |22
               77| 1 |58
             3210| 2 |012
               96| 2 |59

The distributions are fairly similar. Sodium in adult cereals has a median of 205 milligrams, whereas that in child cereals has a median of 190 milligrams. They are similarly spread out. The distribution of sodium in adult cereals is a bit more skewed to the left than is the distribution of sodium in child cereals.

2.135   Political conservatism and liberalism:
   a)   As seen in Example 12, one need not add up every separate number when calculating a mean. This would be unwieldy with the political conservatism and liberalism data. We would have to add up 47 ones, 143 twos, etc. (all the way up to 41 sevens), then divide by the 1,331 people in the study. There's a far easier way. We can find the sum of all values in the study ($\sum x$) by multiplying each possible value (1-7 in this case) by its frequency.

$$\bar{x} = \frac{\sum x}{n} = \frac{47(1) + 143(2) + 159(3) + 522(4) + 209(5) + 210(6) + 41(7)}{1,331} = \frac{5,490}{1,331} = 4.12$$

   b)   The mode, the most common score, is four.
   c)   The median would be the middle score – the 666th score. In this case, that category is four.

2.137   GRE scores:
The best answer is (a).

2.139   Relative GPA:
The best answer is (a). (The standard deviation would allow her to calculate her *z*-score.)

2.141   Bad statistic:
The standard deviation was incorrectly recorded. The standard deviation represents a typical scores distance from the mean. For grades ranging between 26 and 100, a standard deviation of 76 is way too large.

♦♦**2.143 Mean for grouped data:**
In Example 12 and exercise 2.135, the mean could be expressed as a sum. Above, the mean was calculated by multiplying each score by its frequency, then summing these and dividing by the total number of subjects. Alternatively, we could first divide each frequency by the number of subjects, rather than dividing the sum by the number of subjects. Dividing the frequency for a given category by the total number of subjects would give us the proportion. We are just changing the order in which we perform the necessary operations to calculate the mean.

♦♦2.145 Range and standard deviation approximation:
Based on the work of statisticians (the Empirical Rule), we know that most, if not all, data points fall within three standard deviations of the mean if we have a bell-shaped distribution. The formula for this is $\overline{x} = \pm 3s$. If the region from three standard deviations below the mean to three standard deviations above the mean encompasses just about everyone in the data set, we could add the section below the mean ($3s$) to the section above the mean ($3s$) to get everyone in the data set. $3s+3s=6s$. Because the range is defined as everyone in the dataset, we can say that the range is equal, approximately, to $6s$.

♦♦2.147 Using MAD to measure variability:
a) When calculating *MAD*, the value of a given observation is subtracted from the mean, the absolute value is taken, and the resulting number is added to the rest of the deviations before dividing by n (we're taking the mean of these absolute deviations).With greater spread, numbers tend to be further from the mean. Thus, the absolute values of their deviations from the mean would be larger. When we take the average of all these values, the overall *MAD* is larger than with distributions with less spread.
b) We would expect *MAD* to be a little less resistant than the standard deviation. Because we square the deviations, a large deviation has a greater effect.

CHAPTER PROBLEMS: STUDENT ACTIVITIES

The responses for exercises 2.149 and 2.150 will vary for each class depending on the data set used.

⌨2.151 GSS:

Frequency table:

TVHOURS	Count
0	57
1	228
2	231
3	138
4	112
5	46
6	20
7	9
8	26
9	2
10	9
11	1
12	9
13	2
14	2
15	2
16	2
20	3

Histogram:

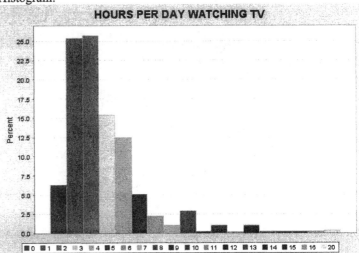

Chapter 3
Association: Contingency, Correlation, and Regression

SECTION 3.1: PRACTICING THE BASICS

3.1 **Which is response / explanatory?:**
  a) The explanatory variable is high school GPA and the response variable is college GPA
  b) The explanatory variable is mother's religion and the response variable is number of children.
  c) The explanatory variable is marital status and the response variable is happiness.

3.3 **Does higher income make you happy?:**
  a) The response variable is happiness and the explanatory variable is income.
  b)

INCOME	HAPPINESS Not Too Happy	Pretty Happy	Very Happy	Total	*n*
Above average	0.08	0.48	0.44	1.00	615
Average	0.09	0.59	0.32	1.00	1420
Below average	0.23	0.57	0.20	1.00	920

People of above average income tend to be the happiest with more people in the above average income bracket responding that they are very happy. People in the below average income bracket are most likely to respond that they are not too happy. Overall, the proportion of people who reported being very happy is 0.31.

3.5 **Alcohol and College Students:**
  a) The response variable is binge drinking and the explanatory variable is gender. We wonder if a person's binge drinking status can be explained in part by their gender. (We don't wonder if a person's gender can be explained by their binge-drinking!)
  b) (i)   There are 1908 male binge drinkers.
      (ii) There are 2854 female binge drinkers.
  c) The counts in (b) cannot be used to answer the question about differences in proportions of male and female students who binge drink. These are not proportions of male and female students; these are counts. There are far more females than males in this study, so it's not surprising that there are more female than male binge drinkers. This doesn't mean that the percentage of women who binge drink is higher than the percentage of men. If we used these numbers, we might erroneously conclude that women are more likely than are men to be binge drinkers.
  d)

Gender	Binge Drinking Status Binge Drinker	Non-Binge Drinker	Total	n
Male	0.49	0.51	1.00	3925
Female	0.41	0.59	1.00	6979

These data tell us that 49% of men are binge drinkers, whereas 51% are not. They also tell us that 41% of women are binge drinkers, whereas 59% are not.
  e) It appears that men are more likely than are women to be binge drinkers.

3.7 **Heaven and hell:**
Since either variable could be considered the outcome of interest, either variable could be taken as the response variable with the remaining being the explanatory variable.

3.9 **Gender gap in party ID:**
  a) The response variable is party identification, and the explanatory variable is gender.
  b) (i) Male and Republican: $399/2771 = 0.14$; (ii) Female and Republican: $422/2771 = 0.15$
  c) (i) Male: $1260/2771 = 0.45$; (ii) Republican: $821/2771 = 0.30$
  d) The figure displays conditional proportions. It suggests that more women than men are Democratic, and that more men than women are Independent or Republican.

SECTION 3.2: PRACTICING THE BASICS

3.11 **Used cars and direction of association:**
a) We would expect a positive association because as cars age, they tend to have covered more miles. Higher numbers on one variable tend to associate with high numbers on the other variable (and low with low).
b) We would expect a negative association because as cars age, they tend to be worth less. High numbers on one variable tend to associate with low numbers on the other.
c) We would expect a positive association because older cars tend to have needed more repairs.
d) We would expect a negative association. Heavier cars tend to travel fewer miles on a gallon of gas.

3.13 **Economic development and air pollution:**
a)

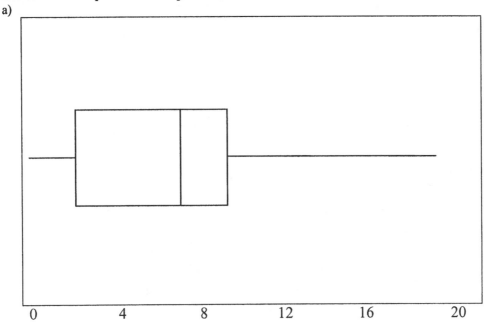

Based on the box plot, I would expect the distribution to be skewed to the right. The maximum is further from the median (and from the nearest quartile) than is the minimum. The standard deviation is almost as big as the mean, and the lowest possible value of 0 is only 6.8/4.7 = 1.45 standard deviations below the mean; these indicate that the distribution is skewed to the right. The standard deviation is likely inflated by the deviation of an extreme positive score. Because the distribution cannot go below 0, we would not expect it to be skewed to the left.
b) The U.S. has both the highest carbon dioxide emissions and the highest GDP with an x coordinate of 34.3 and a y coordinate of 19.7.
c) This is a strong correlation. The closer a correlation is to 1.0 in absolute value, the higher the correlation is.

3.15 **Internet use correlations:**
a) GDP has the strongest linear association with Internet use.
b) Fertility has the weakest linear association with Internet use.
c) This is a negative correlation; thus, as fertility rates increase, Internet use tends to decline. Conversely, as fertility rates decrease, Internet use tends to be greater.

3.17   **What makes $r = 1$?**

a)

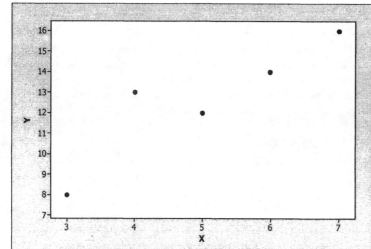

b)   It is the pair: (4, 13).

c)   The value of 13 would have to be changed to 10.

3.19 $r = 0$:

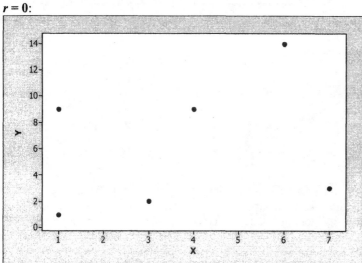

The correlation represented by this scatterplot is about 0.28, but when the data point 1 is removed, it is just about 0.

⌨**3.21** **Which mountain bike to buy?**
   a)   (i) The explanatory variable would be weight, and (ii) the response variable would be price.
   b)

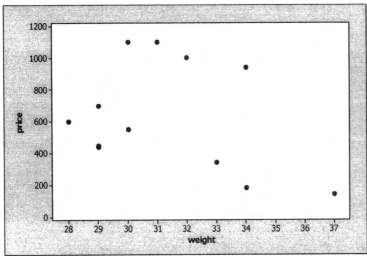

      The relation deviates from linearity in that the bikes with weights in the middle tend to cost the most, with those weighing less and more tending to cost less.
   c)   The correlation is negative and fairly small. This indicates some relation between variables, such that as weight increases, price tends to decrease. Because, however, these variables deviate from linearity in their relation, this correlation coefficient is not an entirely accurate measure of the relation.

▢3.23 Buchanan vote:

a)

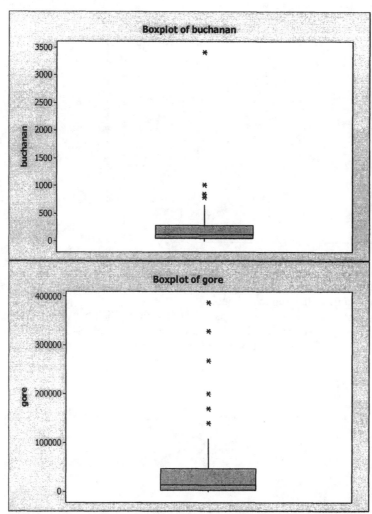

Both box plots indicate that the counts are skewed to the right with few counties in the high ranges of vote counts.

b)

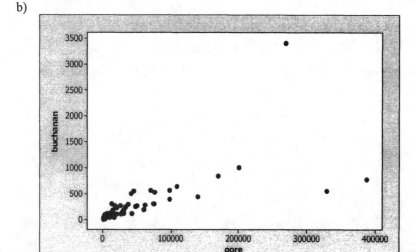

The point close to 3500 on the variable "Buchanan" is a regression outlier; we were unable to make this comparison from the box plots because there were two separate depictions, one for each candidate.

c) We would have expected Buchanan to get around 1000 votes.

d) The box plot for Buchanan would be the same as in part (a).

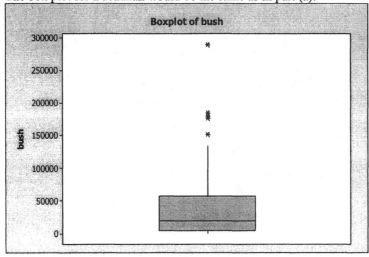

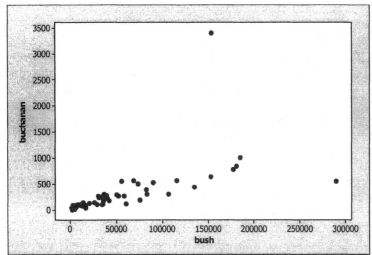

As with the scatterplot with the data for Gore, the point close to 3500 on the variable "Buchanan" is an outlier.

SECTION 3.3: PRACTICING THE BASICS

3.25 **Sit-ups and the 40-yard dash**:
  a)   (i) $\hat{y}$ =6.71-0.024x = 6.71-0.024(10) = 6.47

  (ii) $\hat{y}$ =6.71-0.024x = 6.71-0.024(40) = 5.75

  This is the regression line that would be superimposed on the scatterplot.

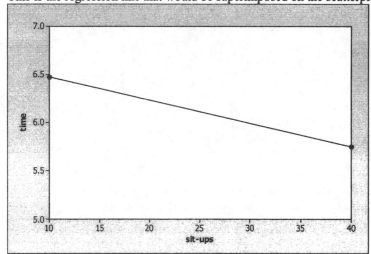

  b)   The y-intercept indicates that when a person cannot do any sit-ups, she/he would be predicted to run the 40-yard dash in 6.71 seconds. The slope indicates that every increase of one sit-up leads to a decrease in predicted running time of 0.024 seconds.
  c)   The slope indicates a negative correlation. The slope and the correlation based on the same data set always have the same sign.

3.27 **Rating restaurants**:
  a)   (i) The predicted cost of a dinner in a restaurant that gets a food quality rating of 0 is \$2.50.
  (ii) The predicted cost of a dinner in a restaurant that gets a food quality rating of 30 is \$122.50.
  b)   For every 1 point increase in food quality rating, the predicted price of the dinner increases by \$4.00.
  c)   The correlation between the cost of a dinner and the food quality rating is 0.53 which is a moderate positive correlation. This indicates that higher costs are associated with restaurants receiving higher food quality ratings.

  d)   The slope can be calculated using the formula $b = r\left(\dfrac{s_y}{s_x}\right) = 0.53\left(\dfrac{20.54}{2.70}\right) = 4.0.$

3.29 **Internet in Ireland**:
  a)   Although slope and correlation usually have different values, they always have the same sign.
  b) $\hat{y}$ =-3.61+1.55x = -3.61+1.55(32.4) = 46.6

  Ireland's predicted Internet use is 46.6%.
  c)   The predicted value is 46.61% (see part c), and the actual value is 23.3% (also see part c).

  The formula for the residual is $y - \hat{y}$. In this case, 23.3-46.6= -23.3.

  The residual is a measure of error; thus, error for this data point is -23.3; it is 23.3% lower than what would be predicted by this equation.

3.31   Air pollution and GDP:

a) (i) $\hat{y} = 1.26+0.346x = 1.26+0.346(0.8) = 1.54$ (rounds to 1.5) per capita carbon dioxide emissions.

(ii) $\hat{y} = 1.26+0.346x = 1.26+0.346(34.3) = 13.13$ (rounds to 13.1) per capita carbon dioxide emissions.

b) $\hat{y} = 1.26+0.346x = 1.26+0.346(34.3) = 13.13$ (rounds to 13.1) per capita carbon dioxide emissions.

The predicted value is 13.1, and the actual value is 19.7.

The formula for the residual is $y - \hat{y}$. In this case, 19.7-13.1 = 6.6.

The residual is a measure of error; thus, error for this data point is 6.6; it is 6.6 higher than what would be predicted by this equation. The U.S. is doing worse than would be predicted in terms of carbon dioxide emissions.

c) $\hat{y} = 1.26+0.346x = 1.26+0.346(28.1) = 10.98$ (rounds to 11.0) per capita carbon dioxide emissions.

The predicted value is 11.0, and the actual value is 5.7.

The formula for the residual is $y - \hat{y}$. In this case, 5.7-11.0 = -5.3.

The residual is a measure of error; thus, error for this data point is-5.3; it is 5.3 lower than what would be predicted by this equation. Switzerland is doing much better than would be predicted with respect to carbon dioxide emissions.

3.33   Regression between cereal sodium and sugar:

a) The software calculates the line for which the sum of squares of the residuals is a minimum. We could approximate this, but would be unlikely to be as exact, and so our line would have larger overall residuals.

b) $\hat{y} = 243.5-0.00708x = 243.5-0.00708(7000) = 193.94$ mg of sodium (rounds to 193.9).

The predicted value is 193.9, and the actual value is 0.

The formula for the residual is $y - \hat{y}$. In this case, 0-193.9 = -193.9.

c) The residual for this point is the lowest along the y-axis, and yes, it does seem to be large relative to the other points on the scatterplot.

d) There is an apparent outlier represented by the leftmost bar. This represents the data point for Frosted Mini Wheats.

3.35   Advertising and sales:

a)

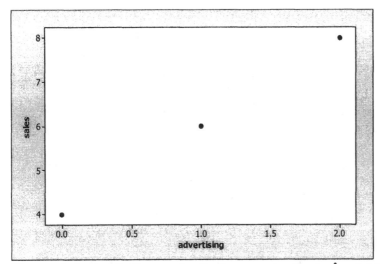

b) The correlation is 1.0. The equation for the regression line is $\hat{y} = 4 + 2x$.

c) **Advertising:** Mean: $\bar{x} = \dfrac{\sum x}{n} = (0+1+2)/3 = 1$;

Standard deviation: $s^2 = \dfrac{\sum (x-\bar{x})^2}{n-1} = ((0\text{-}1)^2 + (1\text{-}1)^2 + (2\text{-}1)^2)/(3\text{-}1) = 2/2 = 1$; $s = \sqrt{s^2} = \sqrt{1} = 1$

Sales: Mean: $\bar{y} = \dfrac{\sum y}{n} = (4+6+8)/3 = 6$;

Standard deviation: $s^2 = \dfrac{\sum (y-\bar{y})^2}{n-1} = ((4\text{-}6)^2 + (6\text{-}6)^2 + (8\text{-}6)^2)/(3\text{-}1) = 8/2 = 4$; $s = \sqrt{s^2} = \sqrt{4} = 2$

d) $b = r\left(s_y / s_x\right) = 1(2/1) = 2$; $a = \bar{y} - b(\bar{x}) = 6 - 2(1) = 4$

$\hat{y} = 4 + 2x$

The y-intercept of 4 indicates that when there is no advertising, it is predicted that sales will be about $4,000. The slope of 2 indicates that for each increase of $1000 in advertising, predicted sales increase by $2000.

3.37 **Predict final exam from midterm:**

a) $b = r\left(s_y / s_x\right) = 0.70(10/10) = 0.70$; $a = \bar{y} - b\bar{x} = 80 - (0.70)(80) = 24$

The regression equation is $\hat{y} = 24 + 0.70x$.

b) The predicted final exam score for a student with an 80 on the midterm is 80.

$\hat{y} = 24 + 0.70(80) = 80$

🖳**3.39** **Study time and college GPA:**

a)

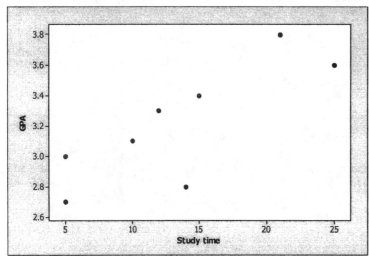

The linear correlation between GPA and study time appears to be positive and fairly strong since the data points follow a positive linear trend.

b) The correlation is 0.81. This indicates that the association between GPA and study time is strong and positive; longer study times are associated with higher GPAs.

c) The prediction equation is $\hat{y} = 2.6252 + 0.0439x$.

(i) a student who studies 5 hours per week is predicted to have a GPA of 2.84

(ii) a student who studies 25 hours per week is predicted to have a GPA of 3.72

⌨**3.41** **Mountain bikes revisited:**
a)

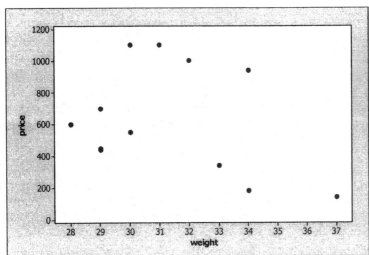

b) From MINITAB, `The regression equation is price = 1896 - 40.5 weight.`
For every 1 unit increase in weight, the predicted price decreases by $40.50. Because it's impossible for a bike to have 0 weight, the y-intercept has no contextual meaning here.

c) $\hat{y} = 1896 - 40.5x = 1896 - 40.5(30) = 681$; the predicted price is $681.

⌨**3.43** **Olympic high jump:**
a) From MINITAB: `The regression equation is Men_Meters = - 8.81 + 0.00560 Year_Men.`
The slope indicates that for each increase of one year, the predicted amount of meters cleared in the high jump by the winning men increased by 0.00560.

b) From MINITAB: `The regression equation is Women_Meters = - 11.5 + 0.00678 Year_Women.`
The slope indicates that for each increase of one year, the predicted amount of meters cleared in the high jump by the winning women increased by 0.0068.

c) Men: $\hat{y} = -8.81 + 0.00560x = -8.81 + 0.00560(2008) = 2.43$

Women: $\hat{y} = -11.5 + 0.00678x = -11.5 + 0.00678(2008) = 2.11$

SECTION 3.4: PRACTICING THE BASICS

3.45 **Men's Olympic long jumps:**
a) The observation in the lower left of the scatterplot may influence the fit of the regression line. This observation was identified because it is a regression outlier.

b) The prediction from the regression line would be more reasonable than would the prediction based on the mean because the regression equation contains more information than the sample mean. For instance, the sample mean does not take into account the positive linear trend exhibited by the data.

c) One would not feel comfortable using the regression line to predict for the year 3000 because that is a huge extrapolation. This data point is too far beyond the data used to determine this line.

3.47 **Murder and education:**
a) $\hat{y} = -3.1 + 0.33(15) = 1.85$

$\hat{y} = -3.1 + 0.33(40) = 10.1$

b) $\hat{y} = 8.0 - 0.14(15) = 5.9$

$\hat{y} = 8.0 - 0.14(40) = 2.4$

c) D.C. is a regression outlier because it is well removed from the trend that the rest of the data follow.

31

d) Because D.C. is so high on both variables, it pulls the line upwards on the right and suggests a positive correlation, when the rest of the data (without D.C.) are negatively correlated. The relationship is best summarized after removing D.C.

3.49 **TV watching and the birth rate**:
a) The U.S. is an outlier on (i) *x,* (ii) *y,* and (iii) relative to the regression line for the other six observations.
b) The two conditions under which a single point can have such a dramatic effect on the slope: (1) the *x* value is relatively low or high compared to the rest of the data; (2) the observation is a regression outlier, falling quite far from the trend that the rest of the data follow. In this case, the observation for the U.S. is very high on *x* compared to the rest of the data. In addition, the observation for the U.S. is a regression outlier, falling far from the trend of the rest of the data. Specifically, TV watching in the U.S. is very high despite the very low birth rate.
c) The association between birth rate and number of televisions is (i) very weak without the U.S. point because the six countries, although they vary in birth rates, all have very few televisions and these amounts don't seem to relate to birth rate. The association is (ii) very strong with the U.S. point because the U.S. is so much higher in number of televisions and so much lower on birth rate that it makes the two variables seem related. A <u>very</u> high number of televisions does coincide with a <u>very</u> low birth rate in the U.S., whereas all the Asian countries are relatively high in birth rates and low in numbers of televisions.
d) The U.S. residual for the line fitted using that point is very small because that point has a large effect on pulling the line downward. There are no other data points near that line, and all other data points are in the far corner, so the line runs almost directly through the U.S. point.

⌨3.51 **Regression between cereal sodium and sugar:**
a)

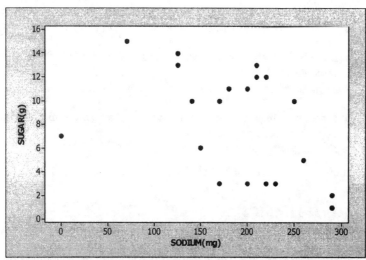

The point at *x*=0 meets the two criteria in that it has an *x* value far from all the others and falls far from the trend that the rest of the data follow.

b) <u>All data points</u>
Regression line: SUGAR(g) = 13.6 - 0.029 SODIUM(mg)
Correlation: -0.45
<u>All except the regression outlier</u>
Regression line: SUGAR(g) = 18.1 - 0.050 SODIUM(mg)
Correlation: -0.62
This point lowers the intercept and makes the slope less steep; overall, the two variables appear less strongly associated when the point is included.

3.53   **Height and vocabulary:**
a)   There is not likely a causal relationship between height and vocabulary. Rather, it is more likely that both increase with age.
b)   Values would be higher on <u>both</u> variables as age increases (e.g., ten-year-olds should be higher on both variables than are five-year-olds). At each age, there should be no overall trend (i.e., some first graders would be high on both, some would be low on both, and some would be low on one and high on the other). Age plays a role in the association because age predicts both height and vocabulary. Height and vocabulary are related because they have a common cause.
c)   In the scatterplot below, we see an overall positive correlation between height (in inches) and vocabulary (assessed on a scale of 1-10) if we ignore grade. However, if we look within each grade, we see roughly a horizontal trend and no particular association. It is age that predicts both height and vocabulary.

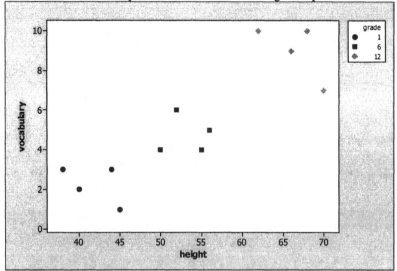

3.55   **Anti-drug campaigns:**
a)   Although there are several possible responses to this exercise, one possible lurking variable could be television watching. Kids who are home watching television are more likely to see these ads and are less likely to be out doing drugs.
b)   Pot smoking (or the lack thereof) might be caused by anti-drug ads, but almost might be caused by other variables, some of which could be associated with anti-drug ads. Including television watching, as mentioned above, other such causal variables could include regular school attendance (a place where students might see more anti-drug ads), parental influence, and neighborhood type.

3.57   **Education causes crime?:**
a) In Minitab, your columns should look similar to the following:

Education	Crime Rate	Rural/Urban
70	140	u
75	120	u
80	110	u
85	105	u
55	50	r
58	40	r
60	30	r
65	25	r

b)

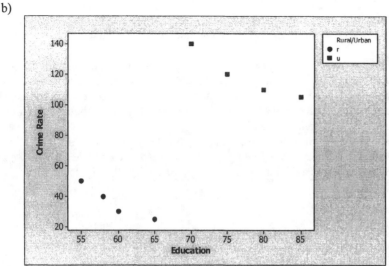

c) The correlation for all 8 data points is 0.73. This indicates a strong, positive linear correlation.

d) (i) the correlation for the urban counties is -0.96 which is a very strong, negative linear correlation; (ii) the correlation for the rural counties is -0.95 which is also a very strong, negative linear correlation. Note that for each subset of data, a higher education rate is associated with a lower crime rate; however, because both the education and crime rates are so much higher for urban counties than for rural, the correlation appears positive when all of the data is considered together. This is a good example of why it is always important to look at a graphical display of your data to determine if a measure of linear correlation is appropriate.

3.59 NAEP scores:
a) The response variable is eighth grade math scores, and the explanatory variable is state.
b) The third variable is race. Nebraska has the overall higher mean because the race ratio is quite different from that in New Jersey. There is a higher percentage of whites and a lower percentage of blacks in Nebraska than in New Jersey, and overall, whites tended to have higher math scores than blacks.

CHAPTER PROBLEMS: PRACTICING THE BASICS

3.61 Choose explanatory and response:
a) The response variable is assessed value, and the explanatory variable is square feet.
b) The response variable is political party, and the explanatory variable is gender.
c) The response variable is income, and the explanatory variable is education.
d) The response variable is pounds lost, and the explanatory variable is type of diet.

3.63 Life after death for males and females:
a)

	Opinion about life after death		
Gender	Yes	No	Total
Male	0.77	0.23	542
Female	0.86	0.14	629

b) Overall, both men and women are more likely to believe in life after death than not, but women are somewhat more likely to do so.

3.65 Degrees and income:
a) The response variable is income. It is quantitative.
b) The explanatory variable is degree. It is categorical.
c) A bar graph could have a separate bar for each degree type. The height of each bar would correspond to mean salary level for a given category.

3.67 **Women managers in the work force:**
a) The response variable is gender, and the explanatory variable is type of occupation.
b) Executive, administrative, and managerial

Year	Female	Male	Total
1972	0.197	0.803	1.00
2002	0.459	0.541	1.00

c) Based on (b), it does seem that there is an association between these variables. Women made up a larger proportion of the executive work force in 2002 than in 1972.
d) The two explanatory variables shown are year and type of occupation.

3.69 **Women in government and economic life:**
a)

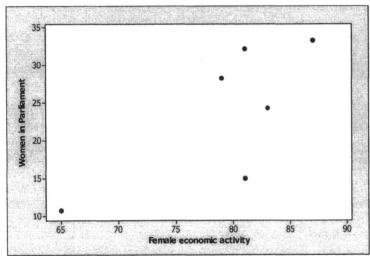

The correlation between women in parliament and female economic activity is 0.745. This correlation is supported by the positive linear trend evident in the scatterplot, but note this is largely driven by the point (for Japan) having female economic activity very low (65).

b) The regression equation is given by $\hat{y} = -48.91 + 0.9186x$. Since the y-intercept would correspond to an x-value of 0, the y-intercept is not meaningful in this case(Female economic activity=0 is outside of the range of observed data).

c) The predicted value for the U.S. is -48.91+0.9186(81)=25.5 with 15.0-25.5= -10.5 as the corresponding residual. The regression equation underestimates the percentage of women in parliament by 10.5% for the U.S.

d) b=0.56(9.8/7.7)=0.7127 and a=26.5-0.7127(76.8)= -28.24. Thus, the prediction equation is given by $\hat{y} = -28.24 + 0.7127x$.

3.71 **Crime rate and urbanization:**
a) An increase of 100 is an increase of 100 times the slope = 0.56(100) = 56. As the urban nature of a county goes from 0 to 100, the predicted crime rate increases by 56%.
b) The correlation indicates a relatively strong, positive relationship.
c) The slope and correlation are related by the formula $b = r\left(s_y / s_x\right)$; $0.56 = 0.67(28.3/34.0)$

3.73 **Height and paycheck:**
a) The response variable is salary, and the explanatory variable is height.
b) The slope of the regression equation is 789 when height is measured in inches and income in dollars. An increase of one inch predicts an increase in salary of $789.
c) An increase of seven inches (from 5 foot 5 to 6 feet) is worth a predicted $789 per inch, or $5523.

3.75 **College GPA = high school GPA**:

The y-intercept would be zero (the line would cross the y-axis at zero when x was zero), and the slope would be one (an increase of one on x would mean an increase of one on y). In this case, the line matches up with the exact points on x and y (0.0 with 0.0, 3.5 with 3.5, etc.). This means that your predicted college GPA equals your high school GPA.

3.77 **Car weight and gas hogs**:

a) The slope indicates the change in y predicted for an increase of one in x. Thus, a 1000 increase in x would mean a predicted change in y of 1000 times the slope: $(-0.0052)(1000) = -5.2$ (poorer mileage).

b) $\hat{y} = 45.6 - 0.0052(6400) = 12.3$. The actual mileage is 17; thus, the residual is $17 - 12.3 = 4.7$. The Hummer gets 4.68 more miles to the gallon than one would predict from this regression equation.

3.79 **Income depends on education?**:

a) For each increase of one percentage in x, we would expect an increase in the predicted value on y by 0.42. Thus, an increase in 10 would be 10 times the slope: $0.42(10) = 4.2$ (or 4200).

b) The slope can be calculated using the formula $b = r(\frac{s_y}{s_x})$. Thus, $0.42 = r(4.69/8.86)$; $r = 0.79$.

(i) The positive sign indicates a positive relationship; as one variable goes up, the other goes up. As one goes down, the other goes down.

(ii) A correlation of 0.79 indicates a strong relationship.

3.81 **Women working and birth rate**:

a) $\hat{y} = 36.3 - 0.30(0) = 36.3$

$\hat{y} = 36.3 - 0.30(100) = 6.3$

When women's economic activity is 0%, predicted birth rate is much higher than when women's economic activity is 100%.

b) The correlation between birth rate and women's economic activity is bigger in magnitude than the correlation between crude birth rate and nation's GNP, indicating a larger association between birth rate and women's economic activity.

3.83 **Income in euros**:

a) The intercept is -20,000 in dollars; thus, the intercept in euros is -16,000.

b) The slope of the regression equation is 4,000 in dollars; thus, the slope in euros is 3,200.

c) The correlation remains the same when income is measured in euros because correlation is not dependent on the units used – whether dollars or euros. It is still 0.50.

3.85 **Murder and single-parent families**:

a) D.C. is the outlier to the far, upper right. This would have an effect on the regression analysis because it is a regression outlier; that is, it is an outlier on x and also is somewhat out of line with the trend of the rest of the data.

b) When D.C. is included, the y-intercept decreases and the slope increases. The D.C. point pulls the regression line upwards on the right side.

⌨3.87 **Violent crime and high school education:**

a)

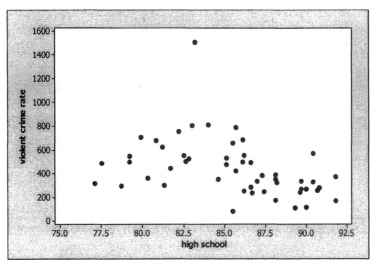

The point with a *y* value of around 1500 is furthest from other data points.

b) `violent crime rate = 2545 - 24.6 high school`

The slope indicates that for each increase of one percent of people with a high school education, the predicted violent crime rate decreases by 24.6.

c) `violent crime rate = 2268 - 21.6 high school`

The slope indicates that for each increase of one percent of people with a high school education, the predicted violent crime rate decreases by 21.6. This is similar to the slope in part b.

⌨3.89 **High school graduation rates and health insurance:**

a)

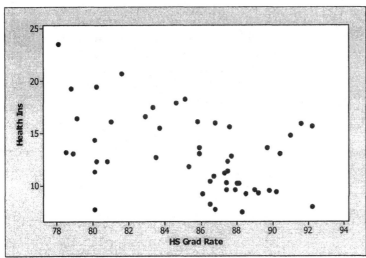

The scatterplot suggests a negative relationship.

b) The correlation is -0.45. As does the scatterplot, this correlation indicates a negative association.

c) The regression equation is

`Health Insurance = 49.2 - 0.42 HS Grad Rate.`

The slope of -0.42 indicates that for each increase of one in the percentage who are high school graduates, the predicted percentage of individuals without health insurance goes down by 0.42. This summarizes the negatively relationship between the variables.

3.91 Income and height:
a) Men tend to be taller and make more money than women. Gender could be the common cause of both of these variables.
b) If gender had actually been measured, it would be a confounding variable. When measured, a lurking variable becomes a confounding variable.

3.93 More sleep causes death?:
a) As people age, they might both sleep more <u>and</u> be more likely to die. This could be the lurking variable that influences both.
b) Subject's age might be the common cause of both of the variables reported in this study. It might actually cause people to sleep more and cause them to be more likely to die.

CHAPTER PROBLEMS: CONCEPTS AND INVESTIGATIONS

⌨3.95 NL baseball team ERA and number of wins:
The responses will be different for each student depending on the methods used.

⌨3.97 Warming in Newnan, GA:
```
The regression equation is Temp = 119 - 0.029 Year
```
The regression line indicates a very slight decrease over time. The Central Park data from Example 12 indicate the opposite, a very slight increase over time.

3.99 Fluoride and AIDS:
San Francisco could be higher than other cities on lots of variables, but that does not
mean those variables cause AIDS, as association does not imply correlation. Alternative explanations are that San Francisco has a relatively high gay population or relatively high intravenous drug use, and AIDS is more common among gays and IV drug users.

3.101 Dogs make you healthier:
Stress level, physical activity, wealth, and social contacts are all possible lurking variables. Any one of these variables may contribute to one's physiological and psychological human health as well as be associated with whether or not a person owns a dog. For example, it may be that people who are more active are more likely to own a dog as well as being physically healthier. Thus, it is possible that one of these lurking variables is responsible for the perceived association between health and dog ownership and if they had been controlled in the study, the association would not be present.

3.103 Properties of r :
The best answer is (b).

3.105 Correct statement about r :
The best answer is (d).

3.107 Slope and correlation:
The best answer is (c).

♦♦3.109 Correlation does not depend on units:
a) If we convert income from British pounds to dollars, then each pound is now worth $2. In other words, we multiply each score by two. Thus, each y-value doubles, the mean of y is now doubled, and the distance of each score from the mean doubles. If this is all so, then the variability, as represented by the standard deviation, has now doubled. As an example, imagine two scores in pounds, 5 pounds and 10 pounds. If both are converted to dollars, they are now $10 and $20. The mean of the two was 7.5 pounds, and it is now $15. The distance of each of the two scores in pounds from the mean was 2.5, but now the distance of each of the two scores in dollars from the mean is $5.
b) The correlation would not change in value, however, because the correlation is based on standardized versions of the measures, and is not affected by the measure used. The formula for the correlation uses z-scores, rather than raw scores. Thus, both pounds and dollars would be converted to z-scores and would lead to the same correlation.

♦♦3.111 Center of the data:
a) Algebraically, we can manipulate the formula $a = \bar{y} - b\bar{x}$ to become $\bar{y} = a + b\bar{x}$. We do this simply by adding $b\bar{x}$ to both sides of the equation. Now $\bar{y}$ is alone on one side of the equation because we have both added and subtracted $b\bar{x}$, canceling each other out. And the other side of the equation is $a + b\bar{x}$. The latter formula is very similar to the regression equation, except the generic predicted y, $\hat{y}$, is replaced by the mean of y, $\bar{y}$. Similarly, the generic x is replaced by $\bar{x}$. Thus, a score on any x that is at the mean will predict the mean for y.
b) Here are the algebraic steps to go from one formula to the other.
Step 1: Because we know that $a = \bar{y} - b\bar{x}$, we can replace the a in the regression equation with this formula. $\hat{y} = \bar{y} - b\bar{x} + bx$
Step 2: We can now subtract $\bar{y}$ from both sides. It cancels itself out on the right, and is now subtracted from the left. $\hat{y} - \bar{y} = -b\bar{x} + bx$ or $\hat{y} - \bar{y} = bx - b\bar{x}$ (if we switch the two parts of the right hand side of the equation)
Step 3: Finally, we can take the b on the right and put it outside of a parenthesis to denote that it is to be multiplied by both variables within the parenthesis. $\hat{y} - \bar{y} = b(x - \bar{x})$
This formula tells us that if we figure out how far from the mean our x is, we can multiple that deviation by the slope to figure out how far from the mean the predicted y is.

CHAPTER PROBLEMS: STUDENT ACTIVITIES

⌨ 3.113 Analyze your data:
The responses to this exercise will vary for each class depending on the data files that each class constructed.

⌨ 3.115 Activity: Guess the correlation and regression:
The responses to this exercise will vary depending on the randomly generated data points.

Chapter 4
Gathering Data

Section 4.1: Practicing the Basics

4.1   Cell phones:
  a)   The response variable was whether a subject had brain cancer. The explanatory variable was cell phone use.
  b)   This was an observational study because the experimenters did not assign subjects to treatments. The researchers observed existing cell phone use.

4.3   Chocolate good for you?:
  a)   The response variable is the death rate due to cancer. The explanatory variable is whether one is a Kuna Indian or a resident of mainland Panama.
  b)   This was an observational study because the rate of death due to cancer was observed (or researched) for the two areas over some period of time. The experimenters did not assign the subjects to treatments.
  c)   It doesn't appear as if the study took lurking variables into account. One possible lurking variable is pollution which could be much higher in mainland Panama and also cause a higher death rate due to cancer.

4.5   School testing for drugs:
Although this study found similar levels of drug use in schools that used drug testing and schools that did not, lurking variables might have affected the results. For example, it is possible that schools that institute drug testing are those in higher crime areas than are those that did not choose to use drug testing. Perhaps the level of drug use in these higher crime communities would have been much higher without these programs than it was with them.

4.7   Hairdressers at risk:
  a)   The response variable is whether or not a woman's infant had a birth defect. The explanatory variable was whether a woman was a hairdresser.
  b)   This study is an observational study. Women were not randomly assigned to be hairdressers or not – they chose this occupation.
  c)   We cannot conclude that there's something connected with being a hairdresser that causes higher birth defect rates. It's possible that there's a lurking variable. For example, the women from the general population might have had higher incomes, and better health insurance. Better neonatal care might have led to the lower rates of birth defects in children of the general population, as compared to children of hairdressers.

4.9   Experiment or observe?:
  a)   Observational study (unethical to assign people to smoking condition)
  b)   Observational study (can't assign people to SAT scores)
  c)   Experiment (can assign recipients to catalog condition)

4.11   Seat belt anecdote:
Anecdotal evidence cannot be expected to be representative of the whole population. The seat belt incident might be the exception, rather than what is typical. Death rates are in fact higher for those who do not wear seat belts.

4.13   Canadian census:
The first census in Canada was conducted in 1665; this was a census of "New France," and it counted 3,215 people. (Note to student: Depending on how you define the first Canadian census, you might come up with a different year and population figure.) The most recent Canadian census in 2001 cited a population size of 30,007,094.

SECTION 4.2: PRACTICING THE BASICS

4.15 **Sample students**:
The first three pairs of two digit numbers are 22, 36, and 84. Because 84 is outside of the range of numbered students (01 to 50), we continue to the next pair: 65. This is also outside of the range, so we continue to the next, 73. This is also outside the range, so we take the next pair, 25. Thus, the three students chosen are those numbered 22, 25, and 36.

4.17 **Auditing accounts - applet**:
Answers will vary each time this is run. What you would do is number the accounts from 01 to 60, then pick random two-digit numbers. You would select the first ten two-digit numbers that fall between 01 and 60. Ignore duplicates.

4.19 **Sampling from a directory**:
First, you would pick five-digit random numbers, ignoring 00000, and numbers above 50,000, as well as duplicates. You would keep the first ten numbers that were in range. Then you would find the names associated with those ten numbers. If, for example, you selected the number 13,050, you would turn to the 131st page (which would include 13,001 to 13,100), and then select the fiftieth name. You would continue until you had ten names.

4.21 **Margin of error and *n***:
a) $\frac{1}{\sqrt{n}} \times 100\% = \frac{1}{\sqrt{100}} \times 100\% = 0.1 \times 100\% = 10$ percentage points, which suggests that between 46% and 46% of Americans believe that New Orleans will never completely recover from Hurricane Katrina.

b) $\frac{1}{\sqrt{n}} \times 100\% = \frac{1}{\sqrt{400}} \times 100\% = 0.05 \times 100\% = 5$ percentage points, which suggests that between 51% and 61% of Americans believe that New Orleans will never completely recover from Hurricane Katrina.

c) $\frac{1}{\sqrt{n}} \times 100\% = \frac{1}{\sqrt{1600}} \times 100\% = 0.025 \times 100\% = 2.5$ percentage points, which suggests that between 53.5% and 58.5% of Americans believe that New Orleans will never completely recover from Hurricane Katrina.

As *n* increases, the sample becomes a more accurate reflection of the population, and the margin of error decreases.

4.23 **Confederates**:
a) This is a leading question because it provides negative information within the question – "symbol of past slavery" and "supported by extremist groups." This negative slant might lead to response bias.
b) This is a better way to ask the question because there is neither explicitly negative nor explicitly positive information about the Confederate symbol within the question.

4.25 **Job market for MBA students**:
a) The population for this survey was all executive recruiters.
b) The intended sample size was 1500, whereas the actual sample size was 97 recruiters. The percentage of nonresponse was 93.53% (rounds to 93.5%).
c) First, it is possible that those who responded were different from those who were recruited to participate, but did not respond. This would constitute a nonresponse bias. For example, recruiters from smaller companies may have had more time to respond; these recruiters may service a different industry (health care, for example) than do recruiters from larger companies. Second, executive recruiters might not be the best source of industries that are hiring. They might, for example, only hear from industries that need help with hiring.

4.27 **Stock market associated with poor mental health:**
a) The population of interest for this survey is Hong Kong residents.
b) This is an observational study because subjects were not assigned to have a particular perception with respect to control over their financial futures. Their existing perceptions were observed.
c) This study used a volunteer sample, the most common kind of convenience sample. Subjects may have chosen to participate because they are more concerned with this topic than are most; for example, these may be people who spend a great deal of time monitoring their financial affairs. In addition, these are people who have access to the Internet. People with internet access are likely to be different on average from other Hong Kong residents. For example, it is likely that they are more educated. It is possible that these problems inflated the relation of perception of control over finances to mental health. Perhaps those who did not participate are less likely to show such a link.

4.29 **Teens buying alcohol over Internet:**
a) This study likely has sampling bias since the sampling method was not random. Not all teenagers are equally likely to respond to an internet survey.
b) It is possible that teenagers who have purchased alcohol over the internet are unlikely to respond to the survey because they are fearful of getting caught. This would introduce nonresponse bias into the study.
c) It is also possible that not all teenagers answer the survey question truthfully, particularly if they are fearful of getting in trouble for answering in the affirmative.

4.31 **Identify the bias:**
a) Undercoverage occurs because not all parts of the Los Angeles population have representation, only those who have subscribed to this newspaper the longest.
b) One problem with sampling design is that the newspaper does not even use random sampling among its subscribers; but instead the 1000 people who have subscribed the longest.
c) Among those who are sent a questionnaire, not all will respond. It is possible that those who respond feel more strongly about the proposal. There was a high percentage of people who did not respond.
d) The question frames the proposal in a positive way, perhaps skewing people's responses in a positive direction – a kind of response bias.

SECTION 4.3: PRACTICING THE BASICS

4.33 **Smoking affects lung cancer?:**
a) This is an experiment because subjects (students in your class) are randomly assigned to treatments (smoking a pack a day, or not ever smoking). Subjects do not choose whether or how much they smoke.
b) One practical difficulty is that it is unethical to assign students to smoke, given that it could cause lung cancer. A second is that we would have no way to ensure that our subjects do as assigned and smoke or not smoke according to the assignment. A third is that we would not have results for fifty years – too long to wait for an answer.

4.35 **More duct tape:**
a) The response variable was whether the wart was successfully removed and the explanatory variable was the type of treatment for removing the wart (duct tape or the placebo). The experimental units were the 103 patients in the Netherlands. The treatments were duct-tape therapy and the placebo.
b) The difference between the number of patients whose warts were successfully removed using the duct tape method and those using the placebo was not large enough to attribute to the treatment type. In other words, the difference in the success rates could be attributed to random variation.

4.37 **Rats and cell phones:**
a) The experimental units are the *n* rats. The explanatory variable is the type of cell phone radiation, and the response variable is whether tumors develop. The treatments are analog cell phone frequency, digital cell phone frequency, and no radiation.
b) The Australian study only considered brain cancer, and the mice were not representative of the overall population of mice; they were susceptible to cancer. It may be that cell phones only cause cancer in those with certain pre-existing characteristics.

4.39   Why randomize:
Randomizing the assignment of treatments helps to eliminate the effects of possible lurking variables so that any observed differences can be attributed to the treatments.

4.41   Blind study:
Subjects should be blind to treatment so that they don't intentionally or unintentionally respond in the way they think they should respond to a given treatment. In this way, subjects are treated as equally as possible.

4.43   Reducing high blood pressure:
a)   We could design an experiment by recruiting volunteers with a history of blood pressure; these volunteers would be the experimental units. The volunteers could be randomly assigned to one of two treatments: the new drug or the current drug. In this experiment, the explanatory variable would be treatment type and the response variable would be blood pressure after the experimental period.
b)   To make the study double-blind, the two drugs would have to look identical so that neither the subjects nor the experimenters who have contact with subjects know what drug a particular subject is taking.

SECTION 4.4: PRACTICING THE BASICS

4.45   Club officers again:
a)   This sample is drawn by numbering the students from 1 to 5. Then, one-digit numbers are randomly picked, and the students selected are the first female student to have her number picked (1 to 3) and the first male student to have his number picked (4 or 5). Numbers beyond the range of 1 to 5 are ignored, as are duplicates. In addition, once a student has been chosen, we will ignore the numbers of other students of that gender. For example, if 2 is picked, we would then ignore 1 and 3, because the other student must be male.
b)   This is not a simple random sample because every sample of size two does not have an equal chance of being selected. For example, samples of two women or two men are prohibited from being selected at all. Moreover, this means that each of the two men has a higher chance of being selected than does any of the three women.

4.47   More school accounts to audit:
a)   Because of the randomness, any number of the accounts sampled could be small. In simple random sampling, every possible sample of a given size has an equal chance of being selected.
b)   First, accounts are labeled 01 to 60. We then select five accounts from 01 to 30 and five from 31 to 60. As we pick random two-digit numbers, we ignore those outside of our range, as well as duplicates. In addition, once we have five from a given stratum – five small accounts, for example – we only choose accounts from the other stratum.
c)   The sample in part b) was a stratified sample since the accounts were formed into two groups (large and small accounts) and then simple random samples of size 5 were taken from each group.

4.49   Smoking and lung cancer:
It is possible that people with lung cancer had a different diet than did those without. For example, these people might have eaten out at restaurants quite a bit, thus consuming more fat. The social aspect of eating out might also have made them more likely to smoke. However, it could have been the fat and not the smoking that caused lung cancer.

4.51   Prospective vs. Retrospective:
In many medical studies, it can be too time consuming and costly to observe patients who exhibit the quality under study until one of the outcomes under study (often death) occurs. It is usually easier, more cost effective and quicker to obtain a sample of subjects who exhibit the quality under study and look back (retrospectively) over past behaviors, exposures, etc.

4.53   **Caffeine jolt**:
   a)   The response variables were blood pressure, heart rate, and reported stress levels. The explanatory variable was whether caffeine was taken with two treatments: caffeine pill or placebo pill. The experimental units were 47 regular coffee drinkers.
   b)   This is a crossover design because all subjects participate in both treatments on different days.

4.55   **Effect of partner smoking in smoking cessation study**:
   a)   This is not a completely randomized design because the researchers are not randomly assigning subjects to living situation (living with another smoker vs. not living with another smoker).
   b)   The experiment has two blocks: those living with smokers and those not living with smokers.
   c)   This is a randomized block design because randomization of units to treatments occurs within blocks.

CHAPTER PROBLEMS: PRACTICING THE BASICS

4.57   **Observational vs. experimental study**:
   In an observational study, we observe people in the groups they already are in. For example, we might compare cancer rates among smokers and nonsmokers. We do not assign people to smoke or not to smoke; we observe the outcomes of people who already smoke or do not smoke.   In an experiment, we actually assign people to the groups of interest. Although it would be unethical, we could turn the above observational study into an experiment by assigning people either to smoke or not to smoke. We would not allow them to make this choice. (Of course, even if we ignored ethics and did this, our subjects might ignore our instructions!) The major weakness of an observational study is that we can't control (such as by balancing through randomization) other possible factors that might influence the outcome variable. For example, in the smoking study, it could be that smokers also drink more, and drinking causes cancer. With the experiment, we randomly assign people to smoke or not to smoke; thus, we can assume that these groups are similar on a range of variables, including drinking. If the smokers still have higher cancer rates than the nonsmokers, we can assume it's because of smoking, and not because of other associated variables such as drinking.

4.59   **Breaking up is hard on your health**:
   a)   The explanatory variable is marriage status, and the response variable is health.
   b)   This is a non-experimental study because people choose their own marital statuses. They are not assigned to be single, married, divorced, or separated.
   c)   It would not be practically possible to design this study as an experiment because one would not be able to assign people randomly to marital status.

4.61   **The fear of asbestos**:
   The friend should give more weight to the study than to the story, which is just anecdotal evidence. Something can be true on average, and yet there can still be exceptions, such as the teacher the friend knows about. The story of one person is anecdotal and not as strong evidence as a carefully conducted study with a much larger sample size.

4.63   **Sampling your fellow students**:
   a)   This would be an example of bias because some parts of the population are favored over others. Moreover, restricting our range in this way would not allow us to determine an association between these variables.
   b)   There are different ways to select a sample that would yield useful information. Here are two. We could select a simple random sample. We could get a list of all students at the school, number them (e.g., 00001 to 10,000 for a school of ten thousand students), then select 20 random five-digit numbers in this range. Alternately, if we wanted to be sure you had equal numbers of first year students through seniors, you could stratify your sample, dividing the students into these categories, and selecting 5 students from each year.

4.65 **Comparing female and male students**:
   a) The students would be numbered from 0001 to 3500. Then we would choose random four-digit numbers, ignoring those outside the range of 0001 to 3500, and ignoring duplicates. We would select the first three students whose numbers matched these criteria.
   b) No; every sample is not equally likely; any possible sample with more than 40 males or fewer than 40 males has probability 0 of being chosen.
   c) This would be a disproportional stratified random sample. It offers the advantage of having the same numbers of men and women in the study, which would be unlikely if the population had a small proportion of one of these, and this is useful for making comparisons.

4.67 **Smallpox fears**:
   a) The population of interest is American adults. There is the possibility of sampling bias in this case. Americans without telephones would not be contacted; moreover, those who are home less frequently would be less likely to be interviewed.
   b) The margin of error was likely calculated by dividing one by the square root of the sample size of 1000. This would be multiplied by 100 to get the percentage points. This means that the range of believable values for a given percentage goes from 3 percentage points below to 3 percentage points above the reported percentage.

4.69 **Video games mindless?**:
   a) The explanatory variable is history of playing video games, and the response variable is visual skills.
   b) This was an observational study because the men were not randomly assigned to treatment (played video games versus hadn't played); those who already were in these groups were observed.
   c) One possible lurking variable is reaction time. Excellent reaction times might make it easier, and therefore more fun, to play video games, leading young men to be more likely to play. Excellent reaction times also might lead young men to perform better on tasks measuring visual skills. These young men might have performed well on tasks measuring visual skills regardless of whether they played video games.

4.71 **Aspirin prevents heart attacks?**:
   a) The response variable was whether they had a heart attack rate; the explanatory variable was treatment group (aspiring or placebo).
   b) This is an experiment because physicians were randomly assigned to treatment – either aspirin or placebo.
   c) Because the experiment is randomized, we can assume that the groups are fairly balanced with respect to exercise. Each group would have some physicians with low exercise and some with high. On average, they'd be similar.

4.73 **Smoking and heart attacks**:
   a) Among those who were in the aspirin group, 0.493 never smoked, 0.397 smoked in the past, and 0.110 are current smokers. Among those in the placebo group, 0.498 never smoked, 0.391 smoked in the past, and 0.111 currently smoke. These proportions are very similar.
   b) It does seem that the randomization process did a good job in achieving balanced treatment groups in terms of smoking status. Because there are similar proportions of physicians in both groups who report that they have never smoked, used to smoke, or currently smoke, this variable is not likely to be responsible for any differences in heart attack rates. The heart attack response between the two groups should not be systematically influenced in one direction or the other due to smoking status of the physicians.

4.75 **Zyban and Nicotine Patch Study Results:**

a) Nicotine patch only: $1/\sqrt{244} \times 100\% = 0.064 \times 100\% = 6.4$ percentage points

Zyban only: $1/\sqrt{244} \times 100\% = 0.064 \times 100\% = 6.4$ percentage points

Nicotine patch with Zyban: $1/\sqrt{245} \times 100\% = 0.064 \times 100\% = 6.4$ percentage points

Placebo only: $1/\sqrt{160} \times 100\% = 0.079 \times 100\% = 7.9$ percentage points

It is believable that the true abstinence percentage falls anywhere within the range indicated by the margin of error. For example, the range for the nicotine patch only is 10% to 22.8%. These are all believable values for the abstinence percentage of those using the nicotine patch only.

b) Yes, it does seem as if the treatments Zyban only and Placebo only are different. The margin of error for Zyban only indicates that the low end of believable values is 23.9%, whereas the margin of error for Placebo only indicates that the high end of believable values is 23.5%. Because there's no overlap, we can conclude that it's likely that these two percentages are significantly different from one another.

c) No, it does not seem as if the treatments Zyban only and Nicotine patch with Zyban are significantly different. There is substantial overlap between the ranges indicated by the margins of error. The range for Zyban only extends from 23.9% to 36.7%, and the range for Nicotine patch with Zyban extends from 29.1% to 41.9%. Because 29.1% through 36.7% are believable values for both treatments, we cannot conclude that there are different abstinence percentages for these two groups.

d) Using the results of parts (a) – (c), the results of the study suggest that two of the treatments, Zyban only and Nicotine patch with Zyban, led to higher abstinence percentages than did either of the other two treatments, Nicotine only or Placebo only. However, there was not a statistically significant difference between Zyban only and Nicotine patch with Zyban. One possible recommendation may be to use Zyban only as an aid for quitting smoking.

4.77 **Comparing gas brands:**

a) The response variable is the gas mileage. The explanatory variable is the brand of gas. Its treatments are Brand A (the name brand) and Brand B (the independent brand).

b) In a completely randomized design, 10 cars would be randomly assigned to Brand A and 10 cars to Brand B.

c) In a matched-pairs design, each car would be a block. It would first use gas from one brand, and then from the other.

d) A matched-pairs design would reduce the effects of possible lurking variables because the two groups would be identical. With a completely randomized design, it is possible that, just by chance, one group of cars gets better gas mileage to begin with than does the other group.

4.79 **Nursing homes:**

a) The nursing homes are clusters.

b) The sample is not a simple random sample because each possible sample does not have an equal chance. For instance, there is no chance of a sample in which just one person from a particular nursing home is sampled.

4.81 **Hazing:**

This is cluster random sampling. The colleges are the clusters.

4.83 **Cell phone study:**

a) This is a retrospective study because people are identified based on current disease status and then are asked questions about their pasts. The cases were those with brain cancer, and the controls were those who did not have brain cancer.

b) Once the data were collected, the researchers could compare the cell phone use of the two groups to see if the people with brain cancer had used cell phones more than had the people who did not have brain cancer.

c) Because this is an observational study, there could be a lurking variable. People with and without brain cancer might be different, on average, on a variable other than cancer status. Even if cell phone use were found to be higher in those with brain cancer than in those without brain cancer, there could a lurking variable, such as age, that is responsible for that association.

CHAPTER PROBLEMS: CONCEPTS AND INVESTIGATIONS

▫4.85 Cell phone use:
The answers to these questions will be different for each student, depending on the study that each student locates.

▫4.87 Internet poll:
Regardless of the study found, the results should not be trusted due to the volunteer nature of the sample.

4.89 Search for an experimental study:
The answers to these questions will be different for each student, depending on the study that each student locates.

4.91 More poor sampling designs:
a) The principal is attempting to use cluster sampling by listing all of her clusters (first-period classes), and taking a random sample of clusters. However, her sample includes only one cluster. She would need to choose several clusters in order to have something resembling a representative sample.
b) Values might be higher than usual because the days sampled are at the start of the weekend. Sampling just Fridays is an example of sampling bias. She should take a simple random sample of all days in the past year.

4.93 Quota sampling:
This is not a random sampling method. People who approach the street corner are interviewed as they arrive (and as they agree to the interview!). Although researchers strive to obtain data from people from a number of backgrounds, the people within these backgrounds (e.g., Hispanic) who are surveyed on the street corner may not be representative of the general population of that kind of person. Although the quota leads to a diversity of people being surveyed, the choice of a given street corner likely constitutes sampling bias.

4.95 Issues in clinical trials:
a) Randomization is necessary because subjects would choose the treatment in which they have the most faith. Such a study would be a measure of how well a treatment works if patients believe in it, rather than how much a treatment works independent of subjects' beliefs about its efficacy.
b) Patients might be reluctant to be randomly assigned to one of the treatments because they might perceive it as inferior to another treatment. In this case, patients might perceive (even in the absence of the data that this study is trying to collect) that the new treatment will be an improvement, and might be reluctant to participate in the study without the guarantee that they can get that treatment.
c) If the researcher thinks that the new treatment is better than the current standard, he or she might be reluctant to proceed because he or she might feel that all patients should get the new treatment, and not just those randomly assigned to it.

4.97 Is a vaccine effective?:
Because the disease is so rare, it's very unlikely that the 200 people randomly chosen to be in this study would have the disease, whether or not they get the vaccine. It would be more practical to find a certain number of people who already have the rare disease (the cases). We would compare the proportion of these people who had received the vaccine to the proportion in a group of controls who did not have the disease.

4.99 Getting a random sample:
The best answer is (c).

4.101 Opinion and question wording:
The best answer is (a).

4.103 Emotional health survey:
The best answer is (d).

4.105   Effect of response categories:
The best answer is (b).

♦♦4.107 Systematic sampling:
a) Although every subject is equally likely to be chosen – at least before the first subject is chosen – every possible sample of 100 is not equally likely. We would never, for example, have a sample that included subjects whose names were next to each other on the population list.
b) The company would determine the first item using a randomly selected two-digit number between 01 and 50; the item that coincided with that number would be checked, and then every 50th item thereafter also would be checked.

♦♦4.109 Mean family size:
Consider the hint. The mean population family size is 6. If we choose families, the possible values are 2 and 10, each equally likely, and the sampling is not biased. If we choose individuals, the value 2 has probability 2/12 and the value 10 has probability 10/12. We are likely to overestimate the mean size.

CHAPTER PROBLEMS: STUDENT ACTIVITIES

4.111   Munchie Capture-Recapture:
If the estimate is not close, one factor that could be responsible is the fairly small sample size. This could be a problem in real-life applications as well, particularly if the animal species of interest were endangered.

4.113   Activity: Sampling the states
Responses will vary each time this exercise is conducted.

REVIEW PROBLEMS: PRACTICING THE BASICS

R1.1   **Believe in astrology?:**
   a)   The sample of interest is the 1245 subjects who responded to the General Social Survey question on astrology. The population of interest is the American adult public.
   b)   The observed variable is the subject's response about whether astrology has scientific truth, which is categorical.
   c)   The sample proportion who responded 'definitely or probably true' is 651/1245≈0.523.

R1.3   **British opinion about President Bush as leader:**
   a)   This variable is categorical since each observation belongs to one of four possible choices (great leader, reasonably satisfactory leader, pretty poor leader, terrible leader).
   b)   These percentages are statistics since they are calculated from a sample.

R1.5   **Religions:**
   a)

Frequency Table: World religions		
Religion	**Frequency (in billions)**	**Percent**
Christianity	2.1	41.18
Islam	1.3	25.49
Hinduism	0.9	17.65
Confucianism	0.4	7.84
Buddhism	0.4	7.84
Total	**5.1**	**1.00**

   b)

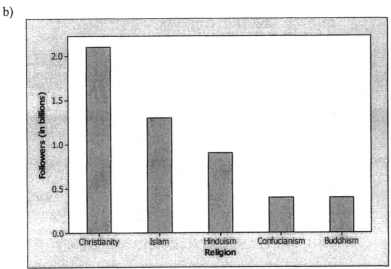

   c)   A mean and median cannot be found for this data because the variable 'religion' is not quantitative and cannot be meaningfully ordered. The modal religion is Christianity.

R1.7   Newspaper reading:
a)   The mode response is the response with the highest frequency, 'every day'. The median response is the one containing the 50th percentile, 'a few times a week'.
b)   The sample mean number of times per week reading a newspaper is calculated as follows:

$$Mean = \frac{945 \cdot 7 + 611 \cdot 3 + 418 \cdot 1 + 404 \cdot 0.5 + 350 \cdot 0}{945 + 611 + 418 + 404 + 350} = 3.3. \cdot$$

The average number of times per week the respondents spent reading a newspaper is 3.3. This is less than the mean of 4.4 for the 1994 GSS, perhaps because of the increased popularity of news sites on the internet.

⌨R1.9 Females in the labor force:
a) $Mean = \dfrac{83 + 82 + 72 + 80 + 80 + 81 + 84 + 81}{8} \approx 80.4$. To find the median, first sort the data: 72, 80, 80, 81, 81, 82, 83, 84. The median is the average of the middle two observations, 81.
b)   In South America, the mean is (48+58+52+50+62+40+51+44+45+68+55)/11≈52.1. Upon comparing the mean values, female economic activity tends to be lower in South America than in Eastern Europe.
c)   Female economic activity is quantitative. Nation is categorical. The response variable is female economic activity.

⌨R1.11 Minimum wage:
a)   The median is found by first sorting the data (5.15, 10, 10.01, 10.25, 10.46) and then finding the data point in the middle position, 10.01. The median minimum wage is $10.01, about half of the minimum wages are less than $10.01 and about half are larger.
The mean is found by summing the data and dividing by the number of data points:
$$\bar{x} = \frac{10 + 10.25 + 10.46 + 10.01 + 5.15}{5} = 9.174.$$ The average minimum wage of the five nations considered is $9.17.
The range is the difference between the largest and smallest observations: 10.46-5.15=$5.31. The difference in the largest and smallest minimum wages for the five nations is $5.31.
In order to find the standard deviation of the observations, the variance is first calculated as follows:
$$s^2 = \frac{\sum (x - \bar{x})^2}{n - 1} = \frac{(10 - 9.174)^2 + (10.25 - 9.174)^2 + (10.46 - 9.174)^2 + (10.01 - 9.174)^2 + (5.15 - 9.174)^2}{4}$$
=5.0963. Thus, $s = \sqrt{5.0963} = 2.26$. A typical minimum wage value is $2.26 from the mean value of $9.17.

b)   The median is found by first sorting the data (10, 10.01, 10.25, 10.46) and then finding the average of the two data points in the middle: (10.01+10.25)/2=$10.13. The median minimum wage is $10.13, about half of the minimum wages are less than $10.13 and about half are larger.
The mean is found by summing the data and dividing by the number of data points:
$$\bar{x} = \frac{10 + 10.25 + 10.46 + 10.01}{4} = 10.18.$$ The average minimum wage of the four nations considered is $10.18.
The range is the difference between the largest and smallest observations: 10.46-10.00=$0.46. The difference in the largest and smallest minimum wages for the four nations is $0.46.
In order to find the standard deviation of the observations, the variance is first calculated as follows:
$$s^2 = \frac{\sum (x - \bar{x})^2}{n - 1} = \frac{(10 - 10.18)^2 + (10.25 - 10.18)^2 + (10.46 - 10.18)^2 + (10.01 - 10.18)^2}{3} = 0.0482.$$
Thus, $s = \sqrt{0.0482} = 0.22$. A typical minimum wage value is $0.22 from the mean value of $10.18. The U.S. minimum wage of $5.15 is an outlier. When it is removed, the mean, range and standard deviation all change quite a bit because they can be largely affected by outliers. The median is not greatly affected by outliers and does not change much when the outlier is removed.

R1.13 Infant mortality:
The infant mortality rates range from 3.0 to 7.0. Since the distances from the median to the minimum and lower quartile are less than the distances from the median to the maximum and upper quartile, the distribution of infant mortality rates is skewed to the right. 25% of the nations had infant mortality rates less than 3.9 and 25% of the nations had infant mortality rates above 6.1.

R1.15 Using water:
a) The most realistic value is 300. -10 is not possible because the standard deviation is always positive. 0 is not possible because the data cover a range of values. Since the data is usually spread between 3 standard deviations of the mean, 10 is too small to be realistic and 1000 is too large. Thus, 300 is the most realistic value for the standard deviation.
b) The most realistic value is 350. The interquartile range covers the middle 50% of the data. It cannot be negative, nor is it 0 for a data set that covers a range of values. Since the median is 500 and the data ranges between 200 and 1700, the interquartile range is unlikely to be either 10 (too small) or 1500 (too large). The most realistic option is 350.

R1.17 Human contacts:
a) Since the mean is quite a bit larger than the median and the distances from the median to the minimum and lower quartile are much smaller than from the median to the maximum and upper quartile, the distribution is skewed to the right.
b)

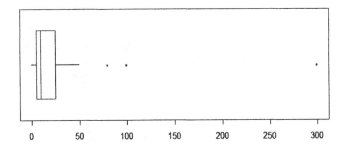

R1.19 Opinion about homosexuality:
a) 416/547 = 0.76, 76%.
b) 213/586 = 0.36, 36%.
c) Yes, whether or not a person believes that homosexual relations are always wrong seems to depend on whether the person considers themselves a liberal or a fundamentalist. The percentage of liberals surveyed who hold this belief is 36% compared to 76% of fundamentalists.

R1.21 How much is a college degree worth?:
a) Using the two points (0, 28645) and (4, 51554) to fit a straight line, we find that
$$slope = \frac{51554 - 28645}{4 - 0} = 5727.25 \cdot$$
b) Age could be a lurking variable, but as it is described here, it would not be responsible for the association. This is because age is positively associated with income but negatively associated with years of education. So, as age increases, income tends to go up and education tends to go down, which has a negative influence on the association between income and education.

R1.23 Child poverty:
a) Since none of the social expenditures as a percent of gross domestic product were lower than 2%, the y-intercept does not have a meaningful interpretation. If 0 were within the range of x-values, 22 would represent the child poverty rate for a country with 0% of their gross domestic product spent on social expenditures. The slope represents the change in y for a one unit change in x. If social expenditure as a percent of gross domestic product increases by 1%, the child poverty rate is predicted to decrease by 1.3%.
b) The predicted poverty rate for the U.S. is 22-1.3(2)=19.4% and the predicted poverty rate for Denmark is 22-1.3(16)=1.2%.

c) As social expenditure increases, the child poverty rate tends to decrease. The association between social expenditure and child poverty rate is strong and negative.

⌨R1.25 U.S. child poverty:
The y-intercept appears to be about 18 and the slope appears to be about 0.5. The estimate of the slope was found using the approximate points (10, 22.5) and (25, 30).

R1.27   Fewer vacations and death:
If higher SES is responsible for both lower mortality and for more frequent vacations, it is likely that including it in the study would remove their apparent association. In other words, the association between vacationing frequency and frequency of deaths from heart attacks would no longer be found significant for those within the same SES.

R1.29   Taxes and global warming:
This is an example of response bias. Although the basis of all three questions is the same, does the respondent favor a tax increase on gasoline, the wording of the last two questions led more of the respondents to answer affirmatively.

Review Problems: Concepts and Investigations

R1.31   Fat, sugar, and health:
a)   slope $\approx$ -0.22
b)   slope $\approx$ 0.23
c)   In part (a), the correlation would be negative, as the amount of fats and sweets eaten increases, diet costs tend to decrease. In part (b) the correlation is positive, as the amount of fruits and vegetables eaten increases, diet costs tend to increase. Also, the correlation has the same sign as the slope.

R1.33   Sneezing at benefits of echinacea:
Anecdotal evidence is usually not representative of the population. In this example, the customers who believed the Echinacea was effective were the ones who were more likely to come back in and tell the manager good things. These customers are unlikely to be representative of the population. In the randomized experiment, the subjects who received Echinacea were randomized so that the effects of potential lurking variables were minimized. The results of such a study are more trustworthy because the results can be attributed to the treatments rather than to lurking variables.

⌨R1.35 Internet time and age:
Answers will vary but should include the following:
The fitted regression equation for 2004 is $\hat{y} = 9.89 - 0.058x$ where y=WWWHR and x=AGE. Thus, for every year increase in age, the number of hours spent per week on the WWW is expected to decrease by 0.058. The correlation between WWWHR and AGE was found to be -0.08 for 2004. This represents a very weak negative association between the two variables.

Chapter 5
Probability in Our Daily Lives

SECTION 5.1: PRACTICING THE BASICS

5.1 **Probability:**
The long run relative frequency definition of probability refers to the probability of a particular outcome as the proportion of times that the outcome would occur in a long run of observations.

5.3 **Vegetarianism:**
No. In the short run, the proportion of a given outcome can fluctuate a lot. Only in the long run does a given proportion approach the actual probability of an outcome.

5.5 **Due for a basket:**
False. For the typical pro, successive shots are independent, and what happens in shots already taken does not affect much, if at all, the chance of making future shots.

5.7 **Polls and sample size:**
No. With a biased sampling design, having a large sample does not remove problems from the sample not being selected to represent the entire population.

5.9 **Life on other planets?:**
We would be relying on our own judgment rather than objective information such as data, and so would be relying on the subjective definition of probability.

5.11 **Unannounced Pop Quiz:**
  a) The results will be different each time this exercise is conducted.
  b) We would expect to get about 50 questions correct simply by guessing.
  c) The results will depend on your answer to part a.
  d) 42% of the answers were "true." We would expect this percentage to be 50%. They are not necessarily identical, because observed percentages of a given outcome can fluctuate in the short run.
  e) There are some groups of answers that appear nonrandom. For example, there are strings of five "trues" and eight "falses," But this can happen by random variation. Typically, the longest strings of trues or falses that students will have will be much shorter than these.

SECTION 5.2: PRACTICING THE BASICS

5.13 **Election study:**
  a)   b)   64 outcomes are possible; $4 \times 4 \times 4 = 64$

5.15   **Pop quiz**:
a)

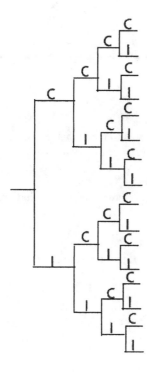

b)   There are 16 possible outcomes ($2 \times 2 \times 2 \times 2$); therefore, the probability of each possible individual outcome is $1/16 = 0.0625$.

c)   Looking at the tree diagram, there are five outcomes in which the student would pass: CCCC, CCCI, CCIC, CICC, and ICCC. The probability of each of these outcomes is 0.0625. Thus, the probability that the student would pass is $0.0625+.0625+.0625+.0625+.0625 = 0.3125$ (rounds to 0.313).

5.17   **Rain tomorrow?**:
False. One can have two outcomes that do not have equal chances of occurring. For example, I might break my leg tomorrow, or I might not. There are two possibilities, but I am happy to know that one is much more likely!

5.19   **All girls in a family**:
a)   The sample space for the possible genders of four children is FFFF, FFFM, FFMF, FFMM, FMFF, FMFM, FMMF, FMMM, MMMM, MMMF, MMFM, MMFF, MFMM, MFMF, MFFM, and MFFF.
b)   There are 16 possible outcomes, only one of which is all girls. The probability of having all girls, therefore, is $1/16 = 0.0625$. Thus, it is fairly unusual to have a family with four children who are all girls.
c)   For the calculation in (b) to be valid, we have to assume that it is equally likely to have a boy or a girl, and that the gender of each child is independent of the genders of the others. That is, we have to assume that the gender of your previous child has no influence on the gender of your next child.

5.21   **Insurance**:
20 heads has probability (1/2) to the 20th power, which is $1/1,048,576 = 0.000001$. The risk of a one in a million death is $1/1,000,000 = 0.000001$.

5.23   **Seat belt use and auto accidents**:
a)   The sample space of possible outcomes is YS; YD; NS; and ND.
b) $P(D) = 2111/577,006 = 0.004$; $P(N) = 164,128/577,006 = 0.284$
c)   The probability that an individual did not wear a seat belt and died is $1601/577,006 = 0.003$. This is the probability that an individual will fall in both of these groups – those who did not wear seat belts and those who died.
d)   If the events N and D were independent, the answer would have been P(N and D) = P(N) × P(D) = $(0.004)(0.284) = 0.001$. In the context of these data, this means that more people died than one would expect if these two events were independent since 0.001 is not equal to 0.003. This indicates that the chance of death depends on seat belt use.

5.25   **Global warming and trees**:

a)

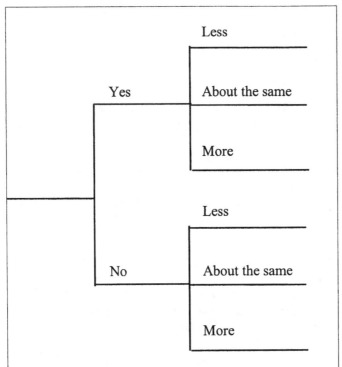

b)   If A and B were independent events, P(A and B)=P(A)×P(B). Since P(A and B) > P(A)P(B), A and B are not independent. Thus, whether or not a person plans to use less fuel in the future depends on whether they believe that global warming is happening. The probability of responding "yes" on global warming and "less" on future fuel use is higher than what is predicted by independence.

5.27   **Arts and crafts sales**:

a)   There are eight possible outcomes as seen in the tree diagram.

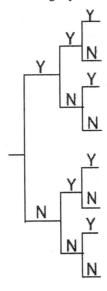

b)   The probability of at least one sale to three customers is 0.488. Of the 8 outcomes, only one includes no sale to any customer. The probability that this would occur = (0.8)(0.8)(0.8) = 0.512. Thus, 1−.512 = 0.488, the probability that at least one customer would buy.

   c) The calculations in (b) assumed that each event was independent. That outcome would be unrealistic if the customers are friends or members of the same family, and encourage each other to buy or not to buy.

SECTION 5.3: PRACTICING THE BASICS

5.29 **Spam:**
   a) $P(B \mid S)$
   b) $P(B^C \mid S^C)$
   c) $P(S^C \mid B^C)$
   d) $P(S \mid B)$

5.31 **Commuting to work:**
   a) The probability that a randomly selected worker commuted to work in a car, truck, or van is $(97,102,050 + 15,634,051)/128,279,228 = 0.879$.
   b) P(car pool | car, truck, or van) = P(car pool and car, truck, or van)/P(car, truck, or van) = 0.122/0.879 = 0.139

5.33 **Revisiting seat belts and auto accidents:**
   a) $P(D) = 2111/577,006 = 0.004$
   b) $P(D \mid$ wore seat belt$) = 510/412,878 = 0.001$
      $P(D \mid$ didn't wear seat belt$) = 1601/164,128 = 0.010$
   c) Neither P(D | wore seat belt) nor P(D | didn't wear seat belt) equals P(D); specifically, 0.001 and 0.010 are different from 0.004. Thus, the events are not independent.

5.35 **Identifying spam:**
   a)

	Identified as Spam by ASG	
Spam	**Yes**	**No**
Yes	7005	835
No	48	

First column total = 7053 = number of messages identified as Spam
First row total = 7840 = number of actual Spam messages
   b) 7005/(7005+835)=0.8935.
   c) 7005/(7005+48)=0.9932.

5.37 **Down syndrome again:**
   a) $P(D \mid NEG) = 6/3927 = 0.0015$
   b) $P(NEG \mid D) = 6/54 = 0.111$; these probabilities are not equal because they are built on different premises. One asks us to determine what proportion of fetuses with negative tests actually has Down Syndrome. This is a small number based on a very large pool of fetuses who had negative tests (3927). The other asks us to calculate what proportion of fetuses with Down syndrome had a negative test. This is a larger number because even though the number of false negatives is small, it's based on a small pool of fetuses (just 54).

5.39 **Happiness in marriage:**
   a) P(very happy) = 861/1410 = 0.61.
   b) (i) P(very happy | male) = 404/637 = 0.63.
      (ii) P(very happy | female) = 457/773 = 0.59.
   c) For these subjects, happiness and being male are not independent. The probability for men to report being very happy is higher than is the overall probability for married adults who report that they are very happy as well as the probability for women to report that they are very happy.

5.41   Shooting free throws:

a)   When two events are not independent: P(makes second and makes first) = P(makes second | makes first) × P(makes first) = (0.60)(0.50) = 0.30

b)   (i)   P(misses second and makes first) = P(misses second | makes first) × P(makes first) = (0.40)(0.50) = 0.20

P(makes second and misses first) = P(makes second | misses first) × P(misses first) = (0.40)(0.50) = 0.20

0.20+0.20 = 0.40 = probability of making one of the two free throws

(ii)   As shown in part a, the probability of making both is 0.30. P(misses second and misses first) = P(misses second | misses first) × P(misses first) = (0.60)(0.50) = 0.30. 0.30+0.30 = 0.60, which is the probability that he makes both or one. Thus, 1-0.60 = 0.40, which is the probability that he makes only one.

c)   The results of the free throws are not independent because the probability that he will make the second shot depends on whether he made the first shot.

5.43   Moral values:

a)   P(B | A)

b)   P(A and B) = P(B | A) × P(A) = (0.80)(0.20) = 0.16

c)   P(A and B) = P(A | B) × P(B)

0.16 = P(A | B) × (0.51)

P(A | B) = 0.16/0.51 = 0.31

5.45   Family with two children:

a)   P(C | A) = P(C and A)/P(A) = (0.25)/(0.5) = 0.5

b)   These are not independent evens because P(C | A), 0.5, is not equal to P(C), 0.25

SECTION 5.4: PRACTICING THE BASICS

5.47   Birthdays of Presidents:

It's easiest to find the probability of no birthday matches among 43 people and then subtract from one. To do this one multiplies (364/365)(363/365)(362/365)...(323/365) = 0.08; 1-0.08 = 0.92. The probability of finding at least one birthday match among 43 people is 0.92. This is not highly coincidental.

5.49   Lots of pairs:

Each student can be matched with 24 other students, for a total of 25(24) pairs. But this considers each pair twice (e.g., student 1 with student 2, and student 2 with student 1), so the answer is 25(24)/2 = 300.

5.51   OSU-Michigan football:

Since there are 10 digits (0 thru 9), each digit has a 0.10 chance of being selected for a given slot. Thus, the probability of any specific 4 number combination, and 4-2-3-9 in particular, is given by $(0.10)^4 = 0.0001$, since this is an intersection of four independent events.

5.53   Coincidence in your life:

The response will be different for each student. The explanation, however, will discuss the context of the huge number of the possible random occurrences that happen in one's life, and the likelihood that at least some will happen (and appear coincidental) just by chance.

5.55   A *true* coincidence at Disneyworld:

a)   The probability that the first will go multiplied by the probability that the second will go and so on for all 5.4 million, that is (1/5000) taken to the 5.4 million power, which is zero to a huge number of decimal places.

b)   This solution assumes that each person decides independently of all others. This is not realistic because families and friends often make vacation plans together.

5.57   **Mammogram diagnostics:**
a)

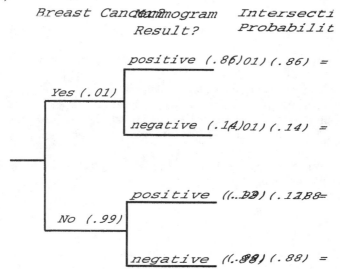

Breast Cancer?   *Mammogram Result?*   *Intersecti Probabilit*

positive (.86) (.01)(.86) =

Yes (.01)

negative (.14) (.01)(.14) =

positive (.12) (.1188=

No (.99)

negative (.88) (.88) =

b)   P(POS) = P(S and POS) + P(S^c and POS) = 0.0086+0.1188 = 0.1274 (rounds to 0.127)
c)   P(S | POS) = P(POS and S)/P(POS) = (0.0086/0.1274) = 0.068
d)   The frequencies on the branches are calculated by multiplying the proportion for each branch by the total number. For example, (0.01)(100) = 1, the frequency on the "yes" branch. Similarly, we can multiply that 1 by 0.86 to get 0.86, which rounds up to 1, the number on the "pos" branch. There are 13 positive tests in this example, only one of which indicates breast cancer. Thus, the proportion of positive tests with breast cancer is 1/(1+12) = 0.077 (rounds to 0.08).

5.59   **Convicted by mistake:**
a)   The chart shows the intersection probabilities. Each is the product of the probability of the outcome in the column and the probability of the outcome in the row.

	Convicted	Acquitted	Total
Guilty	0.855	0.045	0.90
Not guilty	0.005	0.095	0.10
Total	0.86	0.14	1.00

From the bottom of the convicted column, we see that the probability of a defendant being convicted is 0.86. Given that the defendant is convicted, the probability that she or he was actually innocent is P(Innocent | Convicted) = P(Innocent and Convicted)/P(Convicted) = (0.005/0.86) = 0.006.

b)

	Convicted	Acquitted	Total
Guilty	0.475	0.025	0.50
Not guilty	0.025	0.475	0.50
Total	0.50	0.50	1.00

From the bottom of the convicted column, we see that the probability of a defendant being convicted is 0.50. Given that the defendant is convicted, the probability that she or he was actually innocent is P(Innocent | Convicted) = P(Innocent and Convicted)/P(Convicted) = (0.025/0.50) = 0.05.

5.61   **DNA evidence compelling?:**
a)   P(Innocent | Match) = P(Innocent and Match)/P(Match) = 0.0000005/0.4950005 = 0.000001
b)   P(Innocent | Match) = P(Innocent and Match)/P(Match) = 0.00000099/0.00990099 = 0.0001
       When the probability of being innocent is higher, there's a bigger probability of being innocent given a match.
c)   P(Innocent|Match) can be much different from P(Match|Innocent).

⌨5.63 Simulating Donations to Local Blood Bank:
a) The results of this exercise will be different each time it's conducted. One would make the assumption of independence with respect to donors.
b) We would multiple the chances of the first person not being an AB donor (19/20) by the chances of the second person not being an AB donor (19/20) by the chances of the third person not being an AB donor (19/20), etc. This results in multiplying $(0.95) \times (0.95) \times (0.95) \times \cdots \times (0.95)$ – a grand total of twenty 0.95's (or 0.95^{20}). This product of these probabilities is 0.36.

CHAPTER PROBLEMS: PRACTICING THE BASICS

5.65 Due for a boy?:
The gender of each child is independent of the genders of the previous children. Thus, the chance that this child is a boy is still 1/2.

5.67 Choices for lunch:
a) Given that all customers select one dish from each category, there are 18 possible meals [$(2 \times 3 \times 3 \times 1) = 18$].

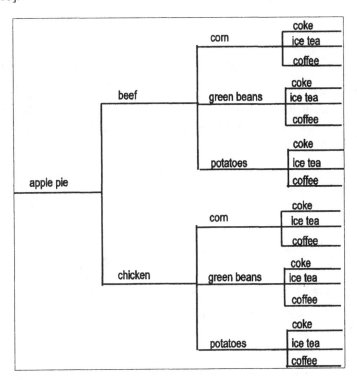

b) In practice, it would not be sensible to treat all the outcomes in the sample space as equally likely for the customer selections we'd observe. Typically, some menu options are more popular than are others.

5.69 Life after death:
a) The estimated probability that a randomly selected adult in the U.S. believes in life after death is 907/1127 = 0.805
b) The probability that both subjects believe in life after death is $0.805 \times 0.805 = 0.65$.
c) The assumption used in answer (b) is that the responses of the two subjects are independent. This is probably unrealistic because married couples share many of the same beliefs.

5.71   Driver's exam:
a)   There are $2 \times 2 \times 2 = 8$ possible outcomes.

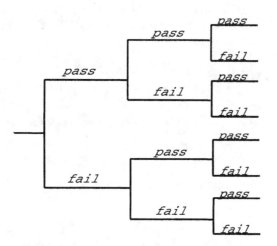

First Friend Second Friend Third F:

b)   If the eight outcomes are equally likely, the probability that all three pass the exam is $1/8 = 0.125$. This could also be calculated by multiplying the probability that the first would pass (0.5), by the probability that the second would pass (0.5), by the probability that the third would pass (0.5). $0.5 \times 0.5 \times 0.5 = 0.125$.

c)   If the three friends were a random sample of their age group, the probability that all three would pass is $0.7 \times 0.7 \times 0.7 = 0.343$.

d)   The probabilities that apply to a random sample are not likely to be valid for a sample of three friends because the three friends are likely to be similar on many characteristics that might affect performance on such a test (e.g., IQ). In addition, it is possible that they studied together.

5.73   Health insurance:
a)   The probability that a patient does not have health insurance is 0.16. Thus, the probability that a patient has health insurance is 1-0.16=0.84.
b)   P(private | health insurance) = P(private and health insurance)/P(health insurance) = 0.59/0.84 = 0.70.

5.75   Teens and parents:
a)   The last two percentages, 31% and 1%, are conditional probabilities. The 31% is conditioned on the event that the teen says that parents are never present during the parties they attend. The 1% is conditioned on the event that the teen says that parents are present at the parties they attend. For both percentages, the event to which the probability refers is a teen reporting that marijuana is available at the parties they attend.
b)

	Parents Present	
Marijuana available	Yes	No
Yes	9	133
No	860	295

a)   The probability that parents are present given that marijuana is not available at the party is P(parents are present and marijuana is not available)/P(marijuana is not available)=860/(860+295) = 0.74.

5.77   Board games and dice:
a)   The sample space of all possible outcomes for the two dice is as follows:
(1,1); (1,2); (1,3); (1,4); (1,5); (1,6); (2,1); (2,2); (2,3); (2,4); (2,5); (2,6);
(3,1); (3,2); (3,3); (3,4); (3,5); (3,6); (4,1); (4,2); (4,3); (4,4); (4,5); (4,6);
(5,1); (5,2); (5,3); (5,4); (5,5); (5,6); (6,1); (6,2); (6,3); (6,4); (6,5); (6,6)
b)   The outcomes in A are: (1,1); (2,2); (3,3); (4,4); (5,5); and (6,6); the probability of this is 6/36 = 0.167.
c)   The outcomes in B are (1,6); (2,5); (3,4); (4,3); (5,2); and (6,1); the probability of this is 6/36 = 0.167.
d)   (i)   There are no outcomes that include both A and B; that is, none of the doubles add up to seven. Thus, the probability of A and B is 0.
       (ii)   The probability of A or B = 0.1667+0.1667 = 0.333.
       (iii) The probability of B given A is 0. If you roll doubles, it cannot add up to seven.
e)   A and B are disjoint. You cannot roll doubles that add up to seven.

5.79 **Conference dinner:**
P(Dinner | Breakfast) = P(Dinner and Breakfast)/P(Breakfast) = 0.40/0.50 = 0.80.

5.81 **A dice game:**
There are 36 possible combinations of dice. Of these, eight add up to seven or eleven [1,6; 2,5; 3,4; 4,3; 5,2; 6,1; 5,6; and 6,5], and four add up to 2, 3, or 12 [1,1; 1,2; 2,1; and 6,6]. Twenty-four add up to other sums. If you roll another sum, it doesn't affect whether you win or lose. Given that you roll a winning or losing combination, there is an 8 in 12 chance of winning. Thus, there is a 0.67 chance of winning. This game would not be played in a casino because the game
favors the player to win in the long run, not the house.

5.83 **Amazing roulette run?:**
a) This strategy is a poor one as the roulette wheel has no memory. The chance of an even slot or of an odd slot is the same on each spin of the wheel.
b) (18/38) to the 26 power, which is essentially 0
c) It would not be surprising if sometime over the past 100 years one of these wheels had 18 evens in a row. Events that seem highly coincidental are often not so unusual when viewed in the context of *all* the possible random occurrences at all times.

5.85 **Screening smokers for lung cancer:**
False negatives would be when the helical computed tomography diagnostic test indicates that an adult smoker does not have lung cancer when he or she does have lung cancer. Conversely, a false positive would occur when this test indicates the presence of lung cancer when there is none.

5.87 **Screening for colorectal cancer:**
a)

b) 15/315 = 0.048 of those who have a positive hemoccult test actually have colorectal cancer. Because so few people have this cancer, most of the positive tests will be false positives. There are so many people without this cancer that even a low false positive rate will result in many false positives.

5.89 **HIV testing**:

a)

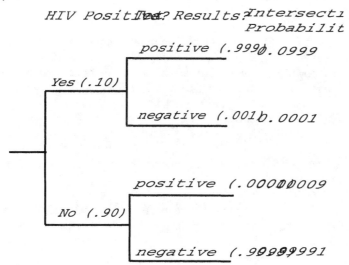

b)

	Positive	**Negative**	**Total**
HIV	0.0999	0.0001	0.1
No HIV	0.00009	0.89991	0.9
Total	0.09999	0.90001	1.0

c) Given that someone has a positive test result, the probability that this person is truly HIV positive is 0.0999/0.09999 = 0.999.

d) A positive result is more likely to be in error when the prevalence is lower as relatively more of the positive results are for people who do not have the condition. With fewer people with HIV, the chances of a false positive are higher. The contingency tables below demonstrate that with a prevalence rate of 10%, there is likely to be one false positive out of 1,000 positive tests, whereas with a prevalence rate of 1%, there is likely to be 1 false positive out of only 101 positive tests.

10%	**Positive**	**Negative**	**Total**
HIV	999	1	1,000
No HIV	1	8,999	9,000
Total	1,000	9,000	10,000

1%	**Positive**	**Negative**	**Total**
HIV	100	0	100
No HIV	1	9,899	9,900
Total	101	9,899	10,000

CHAPTER PROBLEMS: CONCEPTS AND INVESTIGATIONS

⌨**5.91** **Simulate law of large numbers**:

a) The cumulative proportions for parts (i) through (iv) will differ for each student who conducts this exercise. Students will notice, however, that the cumulative proportion of heads approaches 0.50 with larger numbers of flips. This illustrates the law of large numbers and the long-run relative frequency definition of probability in that as the number of trials increases, the proportion of occurrences of any given outcome (in this case, of heads) approaches the actual proportion in the population "in the long run."

b) The outcome will be similar to that in part (a), with the cumulative proportion of heads approaching one third with larger numbers of flips.

5.93 Short term vs. long run:
a) The cumulative proportion of heads would be $60/110 = 0.545$.
b) The cumulative proportion of heads is now $510/1010 = 0.505$.
c) The cumulative proportion of heads is now $5,010/10,010 = 0.500$.
As n increases, the cumulative proportion tends toward 0.50.

5.95 Mrs. Test:
(1) 99% accurate could refer to specificity, meaning that if you are pregnant, you'll have a positive test 99% of the time. (2) It could refer to sensitivity, meaning that if you're not pregnant, you'll have a negative test 99% of the time. (3) 99% chance that you are pregnant given a positive test. (4) 99% chance that you are not pregnant given a negative test.

5.97 Stay in school:
a) Use the extension of the multiplication rule with conditional probabilities.
P(high school) $\times$ (P college | high school) $\times$ P(masters' | high school and college) $\times$ P(Ph.D. | high school, college, and masters') $= (0.80)(0.50)(0.20)(0.30) = 0.024$
b) We multiplied (1) the probability of getting a high school degree by (2) the probability of getting a college degree once you had a high school degree by (3) the probability of getting a masters' degree once you had the earlier degrees by (4) the probability of getting a Ph.D. once you had the earlier degrees.
c) Of those who finish college, 20% get a masters' degree; of these, 30% get a Ph.D.
$(0.20)(0.30) = 0.06$

5.99 Protective bomb:
The fallacy of his logic is that the event of a person bringing a bomb is independent of the event of any other person bringing a bomb. Thus, if the probability of one person bringing a bomb on the plane is one in a million, that is true whether or not this person has a bomb on the plane. The probability of another person bringing a bomb given that this person has a bomb is the same as the probability of another person bringing a bomb given that this person does not have a bomb.

5.101 MC1:
Both (c) and (d) are correct.

5.103 Coin flip:
The best answer is (e).

5.105 Comparable risks:
The best answer is (b).

5.107 TF2:
False. The sample space is TT, HT, TH, and HH. Thus, the probability of 0 heads is ¼, of one head is ½, and ¼ two heads.

5.109 Prosecutor's fallacy:
Being not guilty is a separate event from the event of matching all the characteristics listed. It might be easiest to illustrate with a contingency table. Suppose there are 100000 people in the population. We are given that the probability of a match is 0.001. So out of 100,000 people, 100 people would match all the characteristics. Now the question becomes, of those 100 people, what is the probability that a person would not be guilty of the crime? We don't know that probability. However, let's SUPPOSE in the population of 100,000, 5% of the people could be guilty of such a crime, 95% are not guilty. Here's a contingency table that illustrates this concept.

	Guilty	**Not Guilty**	**Total**
Match		???	100
No Match			99900
Total	5000	95000	100000

Thus, the P(not guilty | match) = ???/ 100.

♦♦ **5.111 Generalizing the multiplication rule:**

We know from this chapter that when two events are not independent, $P(A \text{ and } B) = P(A) \times P(B \mid A)$; if we think about (A and B) as one event, we can see that $P[C \text{ and } (A \text{ and } B)] = P(A \text{ and } B) \times P(C \mid A \text{ and } B)$. If we replace $P(A \text{ and } B)$ with its equivalent, $P(A) \times P(B \mid A)$, we see that:

$P(A \text{ and } B \text{ and } C) = P(A) \times P(B \mid A) \times P(C \mid A \text{ and } B)$.

CHAPTER PROBLEMS: STUDENT ACTIVITIES

⌨**5.113 Simulating matching birthdays:**
   a) The results will be different each time this exercise is conducted.
   b) The simulated probability should be close to 1.

⌨**5.115 Which tennis strategy is better?:**
   a) The results will be different each time this exercise is conducted.
   b) The results will be different each time this exercise is conducted.

Chapter 6
Probability Distributions

Section 6.1 Practicing the Basics

6.1 **Rolling dice**:
a) Uniform distribution; $P(1) = P(2) = P(3) = P(4) = P(5) = P(6) = 1/6$
b) The probabilities below correspond to the stems on the histogram in this exercise. Each probability is calculated by counting how many rolls of the dice add up to a particular number. For example, there are three rolls that add up to four (1,3; 2,2; 3,1) ; thus, the probability of four is $3/36 = 0.083$.

x	$P(x)$
2	1/36
3	2/36
4	3/36
5	4/36
6	5/36
7	6/36
8	5/36
9	4/36
10	3/36
11	2/36
12	1/36

c) The probabilities in (b) satisfy the two conditions for a probability distribution: $0 <= P(x) <= 1$; and
$$\sum P(x) = 1.$$

6.3 **Boston Red Sox hitting**:
a) The probabilities give a legitimate probability distribution because each one is between 0 and 1 and the sum of all of them is 1.
b) $\mu = 0P(0) + 1P(1) + 2P(2) + 3P(3) + 4P(4) = 0(0.718) + 1(0.174) + 2(0.065) + 3(0.004) + 4(0.039) = 0.47$.
   The expected number of bases for a random time at bat for a Boston Red Sox player is 0.47.
c) The mean is the expected value of X; that is, what we expect for the average in a long run of observations. Since the mean is a long term average of values, it doesn't have to be one of the possible values for the random variable.

6.5 **Bilingual Canadians?**:
a) Their corresponding probabilities are not the same; that is, each x value (0, 1, and 2) does not carry the same weight.
b) $\mu = 0(0.02) + 1(0.81) + 2(0.17) = 1.15$

6.7 **Which wager do you prefer?**:
a) There is no correct answer.
b) $\mu = 0.5(200) + 0.5(-50) = 75$
   $\mu = 0.5(350) + 0.5(-100) = 125$
   In this sense, wager two is better.

6.9 **Ideal number of children**:
a) The mean for females is $\mu=0(0)+1(0.03)+2(0.63)+3(0.23)+4(0.11)=2.42$. The mean for males is $\mu=0(0.04)+1(0.03)+2(0.57)+3(0.23)+4(0.13)=2.38$. The means for the two distributions are quite similar.
b) The ideal family size values are a bit more spread out for males, as the smallest value of 0 and largest value of 4 both have higher probabilities for males than females. We would expect the standard deviation to be larger for the males' probability distribution.

c) Although the means are similar, the responses for females tend to be closer to the mean than the responses for males are. Thus, females seem to hold slightly more consistent views than males about ideal family size.

6.11   Selling at the right price:
a)

x	P(x)
$90	0.50
$120	0.20
$130	0.30

$\mu=90(0.50)+120(0.20)+130(0.30)=108$. The expected selling price for the sale of a drill is $108.
b) $\mu=90(0.30)+110(0.40)+130(0.30)=110$. The expected selling price for the sale of a drill under the new pricing strategy is $110. The new strategy will result in higher profits in the long run.

6.13   TV watching:
a) TV watching is, in theory, a continuous random variable because someone could watch exactly one hour of TV or 1.8643 hours of TV.
b) Histograms likely were used because each person reported TV watching by rounding to the nearest whole number.
c) These curves would approximate the two histograms to represent the approximate distributions if we could measure TV watching in a continuous manner. Then, the area above an interval represents the proportion of people whose TV watching fell in that interval.

Section 6.2: Practicing the Basics

6.15   Tail probability in graph:
The observation would fall 0.67 standard deviations above the mean, and thus, would have a *z*-score of 0.67. Looking up this *z*-score in Table A, we see that this corresponds to a cumulative probability of 0.749. If we subtract this from 1.0, we see that the probability that an observation falls above this point (in the shaded region) is 0.251.

6.17   Central probabilities:
a) If we look up 1.64 on Table A, we see that the cumulative probability is 0.9495. The cumulative probability is 0.0505 for -1.64. 0.9495-0.0505 = 0.899, which rounds to 0.90.
b) Using similar logic, we see 0.9951 and 0.0049 on the table for 2.58 and -2.58, respectively. 0.9951-0.0049 = 0.9902, which rounds to 0.99.
c) Finally, we can use the same logic for *z*-scores of 0.67 and -0.67. We find cumulative probabilities of 0.7486 and 0.2514. The difference between these is 0.7486-0.2514 = 0.4972 (rounds to 0.50).
d)

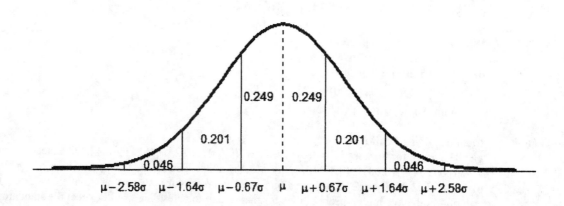

6.19   **Probability in tails for given z-score:**
a)   We have to divide this probability by two to find the amount in each tail, 0.005. We then subtract this from 1.0 to determine the cumulative probability associated with this z-score, 0.995. We can look up this probability on Table A to find the z-score of 2.58.
b)   For both (a) and (b), we divide the probability in half, subtract from one, and look it up on Table A
   (a) 1.96
   (b) 1.64

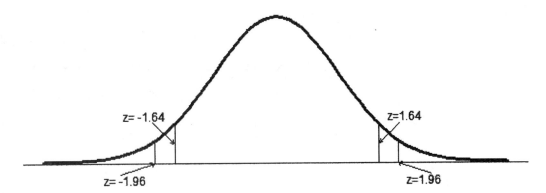

6.21   **z-score and central probability:**
a)   The middle 50% means that 25% of the scores fall between the mean and each z-score (negative and positive). This corresponds to a cumulative probability of 0.75. When looked up in Table A, we find a z-score of 0.67.
b)   Using the same logic as in part (a), we find a cumulative probability of 0.95, and a z-score of 1.64.
c)

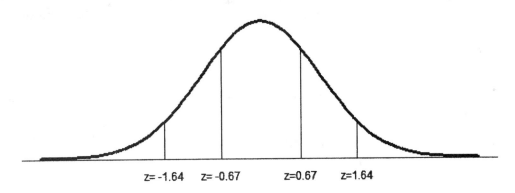

6.23   **Blood pressure:**

a) $z = \dfrac{x - \mu}{\sigma} = (140 - 121)/16 = 1.19$

b)   A z-score of 1.19 corresponds to a cumulative probability of 0.8830. The amount above this z-score would be $1 - 0.88 = 0.12$.

c) $z = \dfrac{x - \mu}{\sigma} = (100 - 121)/16 = -1.31$ which corresponds to a cumulative probability of 0.0951. If we subtract 0.0951 from 0.8830 (the cumulative probability for 140), we get 0.79 as the probability between 100 and 140.

6.25 **Energy use:**

a) $z = \dfrac{x - \mu}{\sigma} = (1000-673)/556 = 0.59$ which corresponds to a cumulative probability of 0.72. Therefore, the probability that household electricity use was greater than 1000 kilowatt-hours is 1-0.72 or 0.28.

b) No, the distribution of energy use doesn't appear to be normal since 0 has a z-score of only $(0-673)/556 = -1.21$, yet energy use cannot be negative.

6.27 **MDI:**

a) (i) $z = \dfrac{x - \mu}{\sigma} = (120 - 100)/16 = 1.25$ which corresponds to a cumulative probability of 0.894.

Subtracted from 1.0, this indicates that the proportion of children with MDI of at least 120 is 0.106.

(ii) $z = \dfrac{x - \mu}{\sigma} = (80 - 100)/16 = -1.25$ which corresponds to a cumulative probability of 0.106.

Subtracted from 1.0, this indicates that the proportion of children with MDI of at least 80 is 0.894.

b) The z-score corresponding to the 99th percentile is 2.33. To find the value of x, we calculate $x = \mu + z\sigma = 100 + (2.33)(16) = 137.3$.

c) The z-score corresponding to the 1st percentile is - 2.33. To find the value of x, we calculate $x = \mu + z\sigma = 100 + (-2.33)(16) = 62.7$.

6.29 **Murder rates:**

a) $z = \dfrac{x - \mu}{\sigma} = (78.5 - 8.7)/10.7 = 6.5$. If the distribution were roughly normal, this would be unusually high.

It is far more than 3 standard deviations above the mean, and most scores in a roughly normal distribution fall within 3 standard deviations of the mean.

b) The standard deviation is higher than the mean, and the lowest possible value of 0 is only $8.7/10.7 = 0.81$ standard deviations below the mean. These are indications that the distribution is skewed to the right.

6.31 **SAT versus ACT:**

$z = \dfrac{x - \mu}{\sigma} = (600 - 500)/100 = 1.0$

$z = \dfrac{x - \mu}{\sigma} = (25 - 21)/4.7 = 0.851$ (rounds to 0.85)

The SAT of 600 is relatively higher than the ACT of 25 because it is further from the mean in terms of the number of standard deviations.

SECTION 6.3: PRACTICING THE BASICS

6.33 **ESP:**
  a) Sample space: (SSS, SSF, SFS, SFF, FSS, FSF, FFS, FFF)
  The probability of each is $(0.5)(0.5)(0.5) = 0.125$. For each number of successes, the probability equals the number of outcomes with that number of successes multiplied by 0.125. For example, there is one outcome with zero successes, giving a probability of 0.125.

x	P(x)
0	0.125
1	0.375
2	0.375
3	0.125

  b) The formula for the binomial distribution is $P(x) = \dfrac{n!}{x!(n-x)!} p^x (1-p)^{n-x}$

$$P(0) = \frac{3!}{0!(3-0)!} 0.5^0 (1-0.5)^{3-0} = (6/6)(1(0.125)) = 0.125$$

$$P(1) = \frac{3!}{1!(3-1)!} 0.5^1 (1-0.5)^{3-1} = (6/2)(0.5(0.25)) = 0.375$$

$$P(2) = \frac{3!}{2!(3-2)!} 0.5^2 (1-0.5)^{3-2} = (6/2)(0.25(0.5)) = 0.375$$

$$P(3) = \frac{3!}{3!(3-3)!} 0.5^3 (1-0.5)^{3-3} = (6/6)(0.125(1)) = 0.125$$

6.35 **Symmetric binomial:**
  a) As calculated in 6.29, the possible values and probabilities would be:

x	P(x)
0	0.125
1	0.375
2	0.375
3	0.125

  This is the graph, which is symmetric.

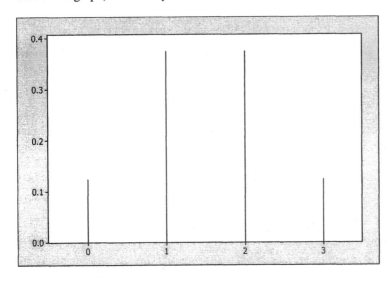

b) As shown in Table 6.5 in the text, the possible values and probabilities would be:

x	P(x)
0	.512
1	.384
2	.096
3	.008

The graph of these probabilities, shown below, is not symmetric.

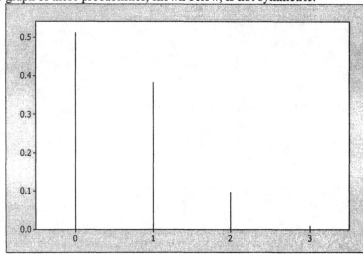

6.37 It's just lunch:

x	P(x)
0	$\binom{2}{0}\left(\frac{7}{8}\right)^2\left(\frac{1}{8}\right)^0 = 0.77$
1	$\binom{2}{1}\left(\frac{7}{8}\right)^1\left(\frac{1}{8}\right)^1 = 0.22$
2	$\binom{2}{2}\left(\frac{7}{8}\right)^0\left(\frac{1}{8}\right)^2 = 0.02$

6.39 NBA shooting:

a) We must assume that the data are binary (which they are – free throw made or missed), that there is the same probability of success for each trial (free throw), and that the trials are independent.

b) $n = 10$; $p = 0.90$

c) (i) $P(x) = \dfrac{n!}{x!(n-x)!}p^x(1-p)^{n-x} = P(10) = \dfrac{10!}{10!(10-10)!}0.9^{10}(1-0.9)^{10-10} = (1)(0.349)(1) = 0.349$

(rounds to 0.35)

(ii) $P(9) = \dfrac{10!}{9!(10-9)!}0.9^9(1-0.9)^{10-9} = (10)(0.3874)(0.1) = 0.3874$ (rounds to 0.39)

6.41 Is the die balanced?:

a) $n = 60$; $p = 1/6 = 0.1667$ (rounds to 0.167)

b) $\mu = np = (60)(0.1667) = 10$

$\sigma = \sqrt{np(1-p)} = \sqrt{(60)(0.1667)(0.8333)} = 2.887$ (rounds to 2.89)

This indicates that if we sample 60 rolls from a population in which 1/6 are 6s, we would expect that about 10 in the sample are 6s. We also would expect a spread for this sample of about 2.887.

c) We would be skeptical because 0 is well over three standard deviations from the mean.

d) $P(0) = \dfrac{60!}{0!(60-0)!} 0.1667^0 (1-0.1667)^{60-0} = (1)(1)(0.0000177) = 0.0000177$

6.43 **Jury Duty:**
a) One can assume that X has a binomial distribution because 1) the data are binary (Hispanic or not), 2) there is the same probability of success for each trial (i.e., 0.40), and 3) each trial is independent of the other trials (whom you pick for the first juror is not likely to affect whom you pick for the other jurors, and $n <$ 10% of the population size). $n = 12$; $p = 0.40$.

b) The probability that no Hispanic is selected is:

$$P(0) = \dfrac{12!}{0!(12-0)!} 0.40^0 (1-0.40)^{12-0} = (1)(1)(0.002) = 0.002$$

c) If no Hispanic is selected out of a sample of size 12, this does cast doubt on whether the sampling was truly random. There is only a 0.2 % chance that this would occur if the selection were done randomly.

6.45 **Checking guidelines:**
a) The sample is less than 10% of the size of the population size; thus the guideline about the relative sizes of the population and the sample was satisfied.

b) The binomial distribution is not likely to have a bell shape because neither the expected number of success (np) nor the expected number of failures [$n(1-p)$] is at least 15. Both are 5.

6.47 **Binomial needs fixed n:**
a) The formula for the probabilities for each possible outcome in a binomial distribution relies on a given number of trials, n.

b) The binomial applies for X. We only have n for X, not Y. $n = 3$. $p = 0.50$.

CHAPTER PROBLEMS: PRACTICING THE BASICS

6.49 **Grandparents:**
a) This refers to a discrete random variable because there can only be whole numbers of grandparents. One can't have 1.78 grandparents.

b) The probabilities satisfy the two conditions for a probability distribution because they each fall between 0 and 1, and the sum of the probabilities of all possible values is 1.

c) The mean of this probability distribution is $0(0.71) + 1(0.15) + 2(0.09) + 3(0.03) + 4(0.02) = 0.50$.

6.51 **More NJ Lottery:**
a) In this lottery, each pick is independent, because digits can occur more than once. Thus, the probability of one of your numbers occurring is 3/10; if the first does occur, the probability of your second number occurring is 2/10; if the first two occur, the probability of your third number occurring is 1/10. Thus, the probability that all three will occur is 0.006. Subtracting from 1.0, we find that the probability that you will not get all three is 0.994.

Winnings	Probability
0	0.994
45.50	0.006

b) The probability that your first number will occur is 1/10, as it is for the second and third. Thus, the probability that all three will occur is 0.001, and the probability that all three will not occur is 0.999.

Winnings	Probability
0	0.999
45.50	0.001

c) The first strategy of picking three different digits is a better strategy because you have a higher chance of winning than with the second strategy.

6.53 **Flyers' insurance:**
a) | Money | Probability |
    |-------|-------------|
    | 0 | 0.999999 |
    | 100,000 | 0.000001 |

b) The mean $= 0(0.999999) + 100,000(0.000001) = 0.10$
c) The company is very likely to make money in the long run because the return on each \$1 spent per flyer averages to \$0.10.

6.55 **z-scores:**
a) First, divide 0.95 in half to determine the amount between the mean and the z-score. This amount is 0.475, which means that 0.975 falls below the z-score in which we're interested. According to Table A, this corresponds to a z-score of 1.96.
b) First, divide 0.99 in half to determine the amount between the mean and the z-score. This amount is 0.495, which means that 0.995 falls below the z-score in which we're interested. According to Table A, this corresponds to a z-score of 2.58.

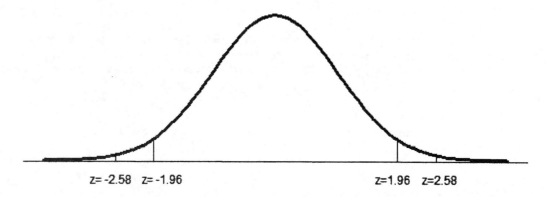

z= -2.58   z= -1.96   z=1.96   z=2.58

6.57 **Quartiles:**
a) If the interval contains 50% of a normal distribution, then there is 25% between the mean and the positive z-score. Added to 50% below the mean, 75% of the normal distribution is below the positive z-score. If we look this up in Table A, we find a z-score of 0.67.
b) The first quartile is at the 25th percentile, and the third is at the 75th percentile. We know that the 75th percentile has a z-score of 0.67. Because the normal distribution is symmetric, the z-score at the 25th percentile must be -0.67.
c) The interquartile range is Q3-Q1 $= (\mu + 0.67\sigma) - (\mu - 0.67\sigma) = 2(0.67)\sigma$.

6.59 **Female height:**
a) $z = (60 - 65)/3.5 = -1.43$; according to Table A, 0.076 women are below this z-score; 0.076 are below five feet.
b) $z = (72 - 65)/3.5 = 2.0$; 0.977 women are below this z-score, which indicates that $1 - 0.977 = 0.023$ are beyond this z-score; that is, over 6 feet.
c) 60 and 70 are equidistant from the mean of 65. If 0.076 are below 60 inches, that indicates that 0.076 are above 70 inches, leaving 0.847 (rounds to 0.85) between them.

6.61 **Gestation times:**
$z = (258 - 281.9)/11.4 = -2.10$; according to Table A, 0.0179 fall below this z-score; thus, proportion 0.018 of babies would be classified as premature.

6.63 **Winter energy use:**
According to Table A, a z-score of 1.5 indicates that 0.93 of the curve falls below this point. Those above this point, $1.0 - 0.93 = 0.07$ households, receive the note.

6.65 **Fast food profits**:

a) $z = (0 - 140)/80 = -1.75$; Table A indicates that 0.0401 (rounds to 0.04) fall below this z-score. The probability that the restaurant loses money on a given day is 0.04.

b) Probability = (0.9599)(0.9599)(0.9599)(0.9599)(0.9599)(0.9599)(0.9599) = 0.7509 (rounds to 0.75). For this calculation to be valid, each day's take must be independent of every other day's take.

6.67 **Manufacturing tennis balls**:

a) $z = (56.7 - 57.6)/0.3 = -3.0$
$z = (58.5 - 57.6)/0.3 = 3.0$
The proportion below a z-score of 3.0 is 0.9987 and below -3.0 is 0.0013. 0.9987-0.0013 = 0.9974 (essentially 1.0) is the probability that a ball manufactured with this machine satisfies the rules.

b) $z = (56.7 - 57.6)/0.6 = -1.5$
$z = (58.5 - 57.6)/0.6 = 1.5$
The proportion below a z-score of 1.5 is 0.933 and below -1.5 is 0.067. 0.933 - 0.067 = 0.866 (rounds to 0.87), which is the probability that a ball manufactured with this machine satisfies the rules.

6.69 **Football wins**:

a) n is 6; p is 0.50; $P(3) = \dfrac{6!}{3!(6-3)!}0.5^3(1-0.5)^{6-3} = 0.31$

b) The assumptions in (a) are that the results of different games are independent, and that the probability of a win is the same in each game. Ordinarily, we would not expect that a team would have exactly the same probability of winning in each game. It is likely, for example, that the outcome of the previous game would affect the outcome of the next.

6.71 **Dating success**:

a) There are three conditions to use the binomial distribution. 1) The outcomes must be binary (e.g., yes or no as the only two options). 2) He must have the same probability of success for each call. 3) Each trial (phone call) must be independent of the others.

b) If he calls the same girl five times, her responses are not likely to be independent of her other responses!

c) $n = 5$; $p = 0.60$; $\mu = 5(0.60) = 3$

6.73 **Female driving accidents**:

a) There are three conditions to use the binomial distribution. 1) The outcomes must be binary (whether or not adult American female dies in a motor vehicle accident); 2) There must be the same probability of success for each person ($p=0.0001$); and 3) The outcome for each woman must be independent of the outcomes for other women. $n = 1,000,000$; $p = 0.0001$.

b) mean = np = (1,000,000)(0.0001) = 100; standard deviation = $\sqrt{np(1-p)} =$
$\sqrt{(1,000,000)0.0001(1-0.0001)} = 9.999$ (rounds to 10.0)

c) The Empirical Rule would suggest that almost all outcomes would fall within three standard deviations of the mean. Thus, this interval would range from 70 to 130.

d) All women would not have the same probability. Outcomes may not be independent; some women may usually ride in the same car as other women. The probability might not be the same for each person; some women may drive more, leading to a higher chance of death in a car accident.

6.75 **Which distribution for sales?**:

a) Since (i) each trial has two outcomes, (ii) the probability of a successful phone call is the same for each call, 2%, and (iii) the trials are independent (the outcome of one call does not affect the outcome of another call) the distribution is binomial.

b) mean= $np = 200(0.02) = 4.0$; standard deviation = $\sqrt{np(1-p)} = \sqrt{200(0.02)(0.98)} = 2.0$. The expected number of successful calls out of 200 is 4.

c) $P(0) = \dbinom{200}{0}(.98)^{200}(.02)^0 = 0.018$.

CHAPTER PROBLEMS: CONCEPTS AND INVESTIGATIONS

6.77 Longest streak made:
a) The mean increases by one for each doubling of number of shots. The chance of a streak becomes longer with more trials.
b) (i) For 400, we would expect 8 because 400 is double 200.
(ii) For 3200, we would expect 11 because we would have to double 400 to 800 to 1600 to 3200.
c) 95% of a bell-shaped curve falls within about two standard deviations of the mean, and 4 is a bit more than two standard deviations.

6.79 Airline overbooking:
a) The sampling distribution of the sample proportion who show up has mean 0.80 and standard error
$\sqrt{p(1-p)/n} = \sqrt{0.8(1-0.8)/190} = 0.029$. Given that almost all sample proportions can be expected to fall within three standard errors of the mean, the airline can expect most flights to have between 0.713 and 0.887 show up. The proportion that the plane can handle is 170/190 = 0.895, just above this range.
b) There might be situations where large groups buy tickets and travel together. This would violate the assumption of independent trials.

6.81 Guess answers:
The best answer is (d).

♦♦6.83 SAT and ethnic groups:
a) A: $z = \dfrac{800-1000}{200} = -1$ and P(Z< 1)=0.16. The proportion not admitted for ethnic group A is 0.16.

B: $z = \dfrac{800-900}{200} = -0.5$ and P(Z<-0.5) =0.31. The proportion not admitted for ethnic group B is 0.31.
b) (0.16+0.31)=0.47 so 0.31/0.41=0.66 of those not admitted are from ethnic group B.
c) A: $z = \dfrac{600-1000}{200} = -2$ and P(Z<-2)=0.023. B: $z = \dfrac{600-900}{200} = -1.5$ and P(Z<-1.5)=0.067. Now, (0.067)/(0.023+0.067)=0.74 are from group B. This problem is worse!

♦♦6.85 Standard deviation of a discrete probability distribution:
$$\sigma^2 = p^2(1-p) + (1-p)^2 p = p(1-p). \text{ So } \sigma = \sqrt{p(1-p)}$$

♦♦6.87 Waiting time for doubles:
a) The probability of rolling doubles is 1/6. Whatever you get on the second die has a 1/6 chance of matching the first. To have doubles occur first on the second roll, you'd have to have no match on the first roll; there's a 5/6 chance of that. You'd then have to have a match on the second roll, and there's a 1/6 chance of that. To get the probability of both occurring is (5/6)(1/6). For no doubles until the third roll, both the first and the second rolls would not match, a 5/6 chance for each, followed by doubles on the third, a 1/6 chance. The probability of all of three events is $(5/6)^2(1/6)$.
b) By the logic in part (a), P(4) = (5/6)(5/6)(5/6)(1/6) or $(5/6)^3(1/6)$. By extension, we could calculate P(x) for any x by $(5/6)^{(x-1)}(1/6)$.

Chapter 7
Sampling Distributions

SECTION 7.1: PRACTICING THE BASICS

7.1 **Simulating the exit poll**:
a) Answers for the sample proportion will vary. Although the population proportion is 0.50, it is unlikely that exactly 50 out of 100 polled voters will vote yes. Sample proportions close to 0.50 will be more likely than those further from 0.50.
b) Simulations will vary. The graph of the sample proportions should be close to bell-shaped and centered around 0.50.
c) Predicted standard error $= \sqrt{\dfrac{0.50(1-0.50)}{100}} = 0.05$.
d) The graph should look similar but shifted so that it is centered around 0.70.

7.3 **Exit poll sampling distribution**:

a) Mean=p=0.10 and standard error$= \sqrt{\dfrac{0.10(1-0.10)}{4000}} = 0.0047$.

b) Mean=p=0.10 and standard error$= \sqrt{\dfrac{0.10(1-0.10)}{1000}} = 0.0095$.

c) Mean=p=0.10 and standard error$= \sqrt{\dfrac{0.10(1-0.10)}{250}} = 0.0190$.

As the sample size gets larger, the standard error gets smaller.

7.5 **Other scenario for exit poll**:
a) The binary variable is whether the voter voted for Schwarzenegger (1) or not (0). For each observation, $P(1) = 0.559$ and $P(0) = 0.441$.
b) Mean $= p = 0.559$

standard error $= \sqrt{\dfrac{p(1-p)}{n}} = \sqrt{\dfrac{0.559(1-0.559)}{2705}} = 0.0095$.

7.7 **Random variability in baseball**:

a) The mean is equal to p or 0.30. The standard error $= \sqrt{\dfrac{p(1-p)}{n}} = \sqrt{\dfrac{0.30(1-0.30)}{500}} = 0.0205$. The shape is approximately normal with a mean of 0.30 and standard deviation of 0.0205.

b) $\dfrac{0.32-0.30}{0.020} = 1.0$ and $\dfrac{0.28-0.30}{0.020} = -1.0$. Since both 0.32 and 0.28 are about a standard deviation from the mean, they would not be considered unusual for this player's year-end batting average

7.9 **Rolling a die:**

a)

Mean	Proportion
1.0	1/36
1.5	2/36
2.0	3/36
2.5	4/36
3.0	5/36
3.5	6/36
4.0	5/36
4.5	4/36
5.0	3/36
5.5	2/36
6.0	1/36

b) The possible outcomes for each die are 1, 2, 3, 4, 5, and 6. Each has an equal chance of occurring (1/6). Thus, the mean is $(1+2+3+4+5+6)/6 = 3.5$.

c) For the sampling distribution listed as the solution in part (a), its symmetric form around the value 3.5 suggests that the mean is 3.5. If we calculate the mean of the sampling distribution in the part (a) solutions, its mean should be 3.5.

7.11 **Syracuse foreign students:**

a) The data distribution consists of 3 1s (denoting a foreign student) and 47 0s (denoting a student from the U.S.).

b) The population distribution consists of the x values of the population of 12,144 full-time undergraduate students at Syracuse University, 4% of which are 1s (denoting a foreign student) and 96% of which are 0s (denoting a student from the U.S.).

c) Mean=p=0.04 and standard error=$\sqrt{\dfrac{0.04(1-0.04)}{50}} = 0.028$. The sampling distribution represents the probability distribution of the sample proportion of foreign students in a random sample of 50 students. In this case, the sampling distribution is approximately normal with a mean of 0.04 and a standard deviation of 0.028.

7.13 **Shapes of distributions:**

a) With random sampling, the data distribution would more closely resemble the population distribution. In both cases, the distributions are based on individual scores, not means of samples.

b) The data distribution would include only 0's and 1's. The sampling distribution of the sample proportion could include any points from 0 to 1. For example, with a sample of 100 observations of which 20 were 0's and 80 were 1's, the proportion would be 0.8.

SECTION 7.2: PRACTICING THE BASICS

7.15   Rolling two dice:
    a)   (i)

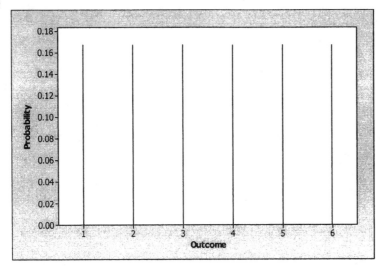

        (ii)

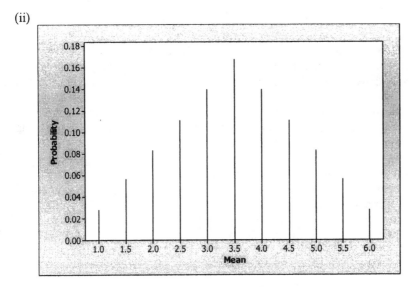

    b)   (i)   The mean for $n=2$ is 3.50, and the standard error is $\sigma/\sqrt{n} = 1.71/\sqrt{2} = 1.21$.

        (ii)   The mean for $n=30$ is 3.50, and the standard error is $\sigma/\sqrt{n} = 1.71/\sqrt{30} = 0.31$

    As n increases, the sampling distribution becomes more normal in shape and less variable.

7.17   Canada lottery:
    a)   The mean would be 0.10, and standard error = $\sigma/\sqrt{n} = 100/\sqrt{1,000,000} = 0.10$

    b)   The z-score at \$1 would be $(1.00-0.10)/0.10 = 9.0$. When we look this up in the Table A, this z-score is not on the table. We find that an area of 0.0 of the curve falls above this z-score. Thus, it's exceedingly unlikely that Joe's average winnings would exceed \$1.

7.19 **Unusual sample mean income?:**

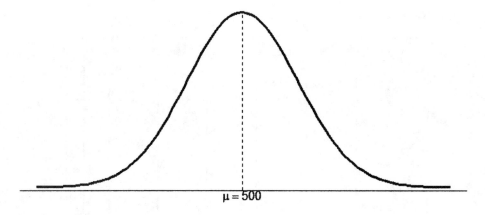

The probability that the sample mean falls above \$540 is

$$P(\overline{X} > 540) = P\left(Z > \frac{540 - 500}{16} \right) = P(Z > 2.5) = 0.006.$$

7.21 **Survey accuracy:**
a) Standard error $= \sigma / \sqrt{n} = 15 / \sqrt{100} = 1.5$

$z = 2/1.5 = 1.33$; the proportion below this z-score is 0.91 (2 is used in the numerator of the z-score equation because it would be the difference between a given sample mean and the population mean no matter what these means actually were), and the proportion below -1.33 is 0.09. Then 0.91 - 0.09 = 0.82 is the proportion that falls within two years of the mean age.

b) If the standard deviation were 10, the standard error would be 1.0. Thus, the sampling distribution of the mean is less variable and the probability of being within two years of the mean age is higher.

7.23 **Household size:**
a) The random variable, $X =$ number of people in a household, is quantitative.
b) The center of the population distribution is the mean of the population of 2.6. The spread is the standard deviation of the population, 1.5.
c) The center of the data distribution is 2.4. The spread is the standard deviation of 1.4.
d) The center of the sampling distribution of the sample mean is 2.6. The spread, or standard error, is

$$\sigma / \sqrt{n} = 1.5 / \sqrt{225} = 0.1$$

⌨**7.25** **CLT for uniform population:**
The results will be different each time this exercise is conducted; however, we should see that the sampling distributions become more bell-shaped and the spread becomes smaller as n increases.

⌨**7.27** **Sampling distribution for normal population:**
The sampling distribution is normal even for $n=2$. If the population distribution is approximately normal, the sampling distribution is approximately normal for all sample sizes.

SECTION 7.3: PRACTICING THE BASICS

7.29   Did Schwarzenegger win?:
The mean of the sampling distribution would be 0.50 and the standard error = $\sqrt{p(1-p)/n}$ = $\sqrt{0.50(1-0.50)/1898}$ = 0.011. We would expect that all sample proportions from this distribution would fall within three standard errors of the mean for a range of 0.47 to 0.53. The proportion who said that they voted for Schwarzenegger was 0.57; as this is outside of the range of expected values, we would have been willing to predict the winner.

7.31   Sampling distributions for the exit poll:
a) n=2705 and p=0.565. Thus, the mean of the sampling distribution of the *number* in the sample who voted for Schwarzenegger is $np \approx 1528$ and the standard deviation is $\sqrt{2705(0.565)(1-0.565)}$ = 25.7841.

b)   The mean of the sampling distribution of the *proportion* of the people in the sample who voted for Schwarzenegger is p=0.565 and the standard deviation is $\sqrt{0.565(1-0.565)/2705}$ = 0.0095.

7.33   Terrorism:
First, they would get the standard error as follows: standard error = $\sqrt{p(1-p)/n}$ = $\sqrt{0.46(1-0.46)/1000}$ = 0.016. If we double this and round it off, we get the margin of error of 0.03.

CHAPTER PROBLEMS: PRACTICING THE BASICS

7.35   Exam performance:
a)   Mean=p=0.70, standard error = $\sqrt{p(1-p)/n}$ = $\sqrt{0.70(1-0.70)/50}$ = 0.0648.

b)   Since n=50, by the Central Limit Theorem we would expect the shape of the sampling distribution to be approximately normal with mean=0.70 and standard deviation = 0.0648.

c)   The z-score for 0.60 is (0.60-0.70)/0.0648=-1.54 giving a cumulative probability of 0.06. It would not be too surprising to only get 60% of the answers correct.

7.37   Alzheimer's:
The shape of the sampling distribution is approximately normal with a mean of p=0.10. (a) For a sample size of 200, the standard error is $\sqrt{0.1(1-0.1)/200}$ = 0.0212. (b) For a sample size of 800, the standard error is $\sqrt{0.1(1-0.1)/800}$ = 0.0106.

7.39   Baseball hitting:
a)   The shape of the sampling distribution is approximately normal with mean=p=0.20 and a standard error= $\sqrt{0.2(1-0.2)/36}$ = 0.0667.

b)   The *z*-score for 0.300 is (0.300 - 0.200)/0.0667=1.50. Since this is well within 3 standard errors of the population proportion, it would not be that surprising for a player to have a batting average of 0.30 after 36 at-bats.

7.41   Aunt Erma's restaurant:
a)   The population distribution has a mean of $900 and a standard deviation of $300.
b)   The data distribution has a mean of $980 and a standard deviation of $276. The standard deviation of the data distribution describes the spread of the daily sales values for this past week.
c)   The mean of the sampling distribution of the sample mean = μ=$900; the standard error= $\sigma/\sqrt{n}$ = $300/\sqrt{7}$ =113.4 dollars. The standard error describes the spread of the sample means based on samples of seven daily sales.

7.43 **Physicians' assistants:**

a) The mean would be \$84,396. The standard error would be $\sigma/\sqrt{n} = 21975/\sqrt{100} = \$2197.5$. The sampling distribution would have a bell shape because the sample size is greater than 30.

b) $z = (80,000 - 84,396)/2197.5 = -2.00$

c) The z-scores are -1.82 and 1.82 enclosing a probability of 0.93.

7.45 **PDI:**

a) $z = (90 - 100)/15 = -0.67$

b) $z = \dfrac{(90 - 100)}{15/\sqrt{225}} = -10.$

c) For an individual PDI value, 90 is only 0.67 standard deviations below the mean and is therefore not surprising. However, it would be unusual for the mean of a sample of size 225 to be 10 standard errors below the mean of its sampling distribution.

♦♦7.47 **Using Control Charts to assess quality:**

a) For a normal distribution, nearly every sample mean will fall within three standard errors of the mean of the sampling distribution. Thus, there is a probability of 0.003 that this process will indicate a problem where none exists.

b) The Empirical Rule says that 95% of the sampling distribution falls within two standard errors of the mean. The probability of falsely indicating a problem would then be 5% if we used two standard errors.

c) (i) The chance that any one sample mean would fall above the mean is 0.50. Thus, the probability that the next 9 means in a row will all fall above the mean is

$$P(9) = \frac{9!}{9!(9-9)!} \, 0.50^9 (1-0.50)^{9-9} = 1(0.00195)(1) = 0.002.$$ (ii) In this scenario, it does not matter

where the first observation falls, only that each succeeding observation falls on the same side as the first. No matter where the first falls, the chance that the next will be on the same side of the mean is 0.50. The probability is twice the probability found in (i), 2(0.002) = 0.004.

CHAPTER PROBLEMS: CONCEPTS AND INVESTIGATIONS

🖳7.49 **CLT for custom population:**
Answers will vary. However, as n increases, the sampling distribution will become closer to normal no matter what distribution the student chose.

7.51 **What good is a standard error?:**
The standard error is the standard deviation of the sampling distribution. It describes how closely a sample statistic will be to the parameter it is designed to estimate, in this case how close the sample proportion will be to the unknown population proportion.

7.53 **Comparing pizza brands:**

a)

Sample	Number Who Prefer Pizza A	Proportion Who Prefer Pizza A
AAA	3	1
AAD	2	2/3
ADA	2	2/3
DAA	2	2/3
ADD	1	1/3
DAD	1	1/3
DDA	1	1/3
DDD	0	0

b) If the proportion of the population who favors Aunt Erma's pizza is 0.50, each of the eight outcomes listed in part (a) are equally likely. Thus, the probability of obtaining a sample proportion of 0 is 1/8, a sample proportion of 1/3 has a (1+1+1)/8=3/8 chance, a sample proportion of 2/3 has a (1+1+1)/8=3/8 chance and a sample proportion of 1 has a 1/8 chance. This describes the sampling distribution of the sample proportion for n=3 with p=0.50.

c) The probability of obtaining a sample proportion of 1/3 is the same as having 1 out of the 3 subjects sampled stating that they prefer Aunt Erma's pizza. $P(1) = \binom{3}{1} 0.5^1 0.5^{3-1} = 0.375 = 3/8$.

7.55 Come out ahead in lottery:
a) If each game costs $1, you'd spend $52 to play 52 times. If a win garners you $5, you'd have to win at least 11 times to earn at least $55 and be ahead of the $52 you spent.
b) The binomial equation allows us to know the probability of a given number of successes, given a certain number of trials and a certain probability. If we were to calculate the probability for 11 wins out of 52 trials, and then again for 12 wins out of 52, 13 wins out of 52, and so on, up to 52 wins out of 52, we'd get the probability that we'd win more than the $52 spent. As the text tells us, this is 0.013.

7.57 Standard error:
The best answer is (b).

7.59 Sampling distribution:
The best answer is (a).

♦♦7.61 Standard error of a proportion:
Standard error = $\sigma / \sqrt{n} = \sqrt{p(1-p)} / \sqrt{n} = \sqrt{p(1-p)/n}$.

CHAPTER PROBLEMS: STUDENT ACTIVITIES

7.63 Simulate a sampling distribution:
c) You expect a mean of about 47.2, which is the population mean.
d) You expect a standard deviation of about 4.9, which is the standard error of the sampling distribution when n = 9.

7.65 Sample vs sampling:
a) The histograms will be different each time this exercise is conducted, but we should have a positively skewed distribution.
b) This is a sampling distribution, and with ten coins in each sample, the distribution should be closer to a normal distribution than that in part (a). It should also be less spread out than the distribution in (a).
This exercise illustrates the floor effect (no coin can have a score below 0, but the scores can go quite high). It also illustrates the Central Limit Theorem in that the increase in sample size makes the distribution somewhat closer to a normal distribution.

REVIEW PROBLEMS: PRACTICING THE BASICS

R2.1 **Vote for Schwarzenegger?:**
  a) $P(A^C)=1-P(A)=1-0.56=0.44$.
  b) The probability for the complement of an event A.

R2.3 **Embryonic stem cell research:**
  a) $P(A \text{ or } B)=P(A)+P(B)=0.03+0.03=0.06$.
  b) The addition rule for the union of two disjoint events.

R2.5 **Heaven and hell:**
  a) $1-0.85=0.15$.
  b) $P(\text{heaven and hell})=P(\text{hell } | \text{ heaven})P(\text{heaven})=0.84(0.85)=0.71$.

R2.7 **UK lottery:**
  a) 6/49.
  b) $np=2600/13,983,816=0.000186$.
  c) You would need to play 268,920 years to expect to win once.

R2.9 **Fraternal bias?:**

$$P(0) = \binom{4}{0}(0.80)^0(0.20)^4 = 0.0016\cdot$$

R2.11 **Quantitative GRE scores:**
  a) (i) $z = \dfrac{700-591}{148} = 0.74$ which has a cumulative probability of 0.77. (ii) 1-0.77=0.23.
  b) The distance between the mean and lowest score is much greater than between the mean and highest score. Also, the highest score is only about 1.4 standard deviations above the mean.
  c) The sampling distribution of the sample mean quantitative exam score for a random sample of 100 people who take this exam is approximately normal with mean=591 and standard error = $148/\sqrt{100} =14.8$.

R2.13 **Spring break drinking:**
  a) standard error= $\sqrt{\dfrac{p(1-p)}{n}} = \sqrt{\dfrac{0.74(1-0.74)}{644}} = 0.017$.
  b) The standard error describes how much we can expect the sample proportion to vary from one sample of size 644 to the next.

R2.15 **Ice cream sales:**
  a) For the population distribution, mean=1000 and standard deviation=300.
  b) For the data distribution, mean=880 and standard deviation=276.
  c) For the sampling distribution of the sample mean for a random sample of 7 daily sales, mean=1000 and standard error= $\sigma/\sqrt{n} = 300/\sqrt{7} =113.$. The standard error describes how much we can expect the sample mean to vary from one sample of 7 daily sales to the next.

R2.17   NY exit poll:
Assuming that the proportion voting for Hillary Clinton were 0.50, the sampling distribution of the sample proportion who voted for Hillary Clinton would be approximately normal with mean=0.50 and standard error= $\sqrt{\dfrac{0.5(1-0.5)}{1336}} = 0.0137$. The z-score associated with a sample proportion value of 0.67 would then be (0.67-0.5)/0.0137=12.4. Since the associated probability is essentially 0, we would conclude that the population proportion of those voting for Hillary Clinton is in fact larger than 0.50 and predict Hillary Clinton to be the winner of the election.

REVIEW PROBLEMS: CONCEPTS AND INVESTIGATIONS

R2.19   Breast cancer gene test:
a)   P(B|H)=0.25.
b)   P(B^C|H^C)=0.95.
c)

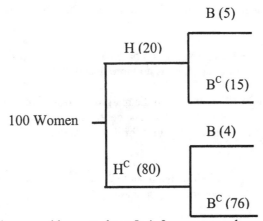

Total, we would expect about 5+4=9 women to relapse.

R2.21   Sample means vary:
Since samples are merely representative of the population, not the same as the population, we would not expect to obtain the exact same sample mean each time a sample of size n is collected. Rather, we would expect these values to vary about the true population mean. The amount by which the sample means vary is summarized by the sampling distribution of the sample mean which has mean=μ and standard error = $\sigma/\sqrt{n}$ where μ is the population mean, σ is the population standard deviation and n is the sample size.

R2.23   Data and population:
True.

R2.25   CLT:
True.

Chapter 8
Statistical Inference: Confidence Intervals

SECTION 8.1: PRACTICING THE BASICS

8.1 **Health care:**
   a) This study will estimate the population proportion who have health insurance and the mean dollar amount spent on health insurance by the population.
   b) We can use the sample proportion and the sample mean to estimate these parameters.

8.3 **Believe in hell?:**
   The point estimate is 55.4%, or 0.554.

8.5 **Watching TV:**
   a) The sample point estimate is the mean of these responses, 2.0.
   b) The margin of error indicates that the sample mean likely falls within 0.91 of the population mean.

8.7 **Believe in heaven?:**
   a) The margin of error for a 95% confidence interval would be 1.96 times the standard error of 0.01, which is 0.02. It is very likely that the population proportion is no more than 0.02 lower or 0.02 higher than the reported sample proportion.
   b) The 95% confidence interval includes all points within the margin of error of the mean. Lower endpoint: $0.86 - 0.02 = 0.84$; upper endpoint: $0.86 + 0.02 = 0.88$. The confidence interval goes from 0.84 to 0.88. This is the interval containing the most believable values for the parameter.

8.9 **CI for loneliness:**
   The confidence interval would range from 0.12 below the mean to 0.12 above the mean: 1.38 to 1.62. This is the range that includes the most believable values for the population mean.

SECTION 8.2: PRACTICING THE BASICS

8.11 **Drug abuse:**
   Margins of error reported in the media typically are based on a 95% confidence interval. They would have multiplied 1.96, the z-score for a 95% confidence interval, by the standard error $= \sqrt{.29(1-.29)/1000}$ giving a margin of error of 0.03 on the proportion scale, or 3%.

8.13 **How green are you?:**
   a) The point estimate is $344/1170 = 0.294$.
   b) $se = \sqrt{\hat{p}(1-\hat{p})/n} = \sqrt{0.294(1-0.294)/1170} = 0.013$. The margin of error is $(1.96)(0.013) = 0.025$.
   c) $0.294 - 0.025 = 0.27$
       $0.294 + 0.025 = 0.32$
       The numbers represent the most believable values for the population proportion.
   d) We must assume that the data are obtained randomly, and that a large enough sample size is used so that the number of successes and the number of failures both are greater than 15. Both seem to hold true in this case.

8.15 **Favor death penalty:**
   a) We can obtain the value reported under "Sample p" by dividing the number of those in favor by the total number of respondents, 1885/2815.
   b) We can be 95% confident that the proportion of the population who are in favor of the death penalty is between 0.652 and 0.687, or rounding, (0.65, 0.69).

c) 95% confidence refers to a probability that applies to the confidence interval method. If we use this method over and over for numerous samples, in the long run we make correct inferences (that is, the confidence interval contains the parameter) 95% of the time.

d) We can conclude that more than half of all American adults were in favor because all the values in the confidence interval are above 0.50.

8.17 Reincarnation:

a) The "sample p" is the proportion of all respondents who believe in reincarnation, $594/2201 = 0.27$. The "95% CI" is the 95% confidence interval. We can be 95% confident that the population proportion falls between 0.25 and 0.29.

b) The sample proportion who did not believe in reincarnation is found by calculating $1 - \hat{p} = 1 - 0.27 = 0.73$. Subtracting each endpoint from 1 gives (0.71, 0.75).

8.19 Instant messaging:

We can be 95% confident that the population proportion of teens who regularly use instant messaging is between 0.72 and 0.78.

8.21 Fear of breast cancer:

a) The proportion is 0.61, and the standard error is $\sqrt{\hat{p}(1-\hat{p})/n} = \sqrt{0.61(1-0.61)/1000} = 0.015$. The 90% confidence interval uses $z = 1.645$. The margin of error is $(1.645)(0.015) = 0.0247$, giving a confidence interval of 0.585 to 0.635. We can be 95% confident that the population proportion falls within this range.

b) For the inference to be valid, the data must be obtained randomly. In addition, the number of successes and the number of failures both must be greater than 15 (they are).

8.23 Exit poll predictions:

a) Sample proportion $= 660/1400 = 0.471$
$$se = \sqrt{\hat{p}(1-\hat{p})/n} = \sqrt{0.471(1-0.471)/1400} = 0.013.$$
margin of error $= (1.96)(0.013) = 0.025$
confidence interval: 0.446 to 0.496
We could predict the winner because 0.50 falls outside of the confidence interval. It does not appear that the Democrats received more than half of the votes.

b) The margin of error for a 99% confidence interval is $(2.58)(0.013) = 0.034$. The confidence interval is 0.437 to 0.505. We now cannot predict a winner because it is plausible that the Democrats received more than half of the votes. The more confident we are, the wider the confidence interval.

8.25 Simulating confidence intervals:

The results of the simulation will be different each time it is conducted. The percentage we'd expect would be 95% and 99%, but the actual values may differ a bit because of sampling variability .

SECTION 8.3: PRACTICING THE BASICS

8.27 Females' ideal number of children:

a) The point estimate of the population mean is 3.16.

b) The standard error of the sample mean: $se = s/\sqrt{n} = 1.91/\sqrt{1097} = 0.0577$.

c) We're 95% confident that the population mean falls between 3.05 and 3.28.

d) It is not plausible that the population mean is 2 because it falls outside the confidence interval.

8.29 Using *t* table:

a) 2.776

b) 2.145

c) 2.977

8.31   **eBay without buy-it-now option:**
a)   First, the data production must have used randomization. Second, the population distribution is assumed to be approximately normal, although this method is robust except in cases such as extreme skew or severe outliers. Here the method is questionable because of an outlier at 178.

b)   The 95% confidence interval would include the sample mean $\pm\ t_{.025}\ (se)$

$se = s/\sqrt{n} = 21.94/\sqrt{18} = 5.1713$
$231.61 - (2.110)(5.1713) = 220.70$
$231.61 + (2.110)(5.1713) = 242.52$
We can be 95% confident that the mean population price is between \$220.70 and \$242.52.

c)   IQR = 246.75 − 221.25 = 25.5; 1.5 X IQR = (1.5)(25.5) = 38.25
Q1 − IQR = 221.25 − 38.25 = 183
Q3 + IQR = 246.75 + 38.25 = 285
There is one potential outlier according to this criterion: 178.

⌨d)   From MINITAB: Variable   Mean StDev
                                 Selling Prices   234.76   17.92

$se = s/\sqrt{n} = 17.92/\sqrt{17} = 4.346$
$234.76 - (2.120)(4.3462) = 225.55$
$234.76 + (2.120)(4.3462) = 243.97$
This confidence interval is narrower than the one using all of the data.

8.33   **TV watching for Muslims:**
a)   The data must be produced randomly, and the population distribution should be approximately normal.
b)   The 95% confidence interval indicates that the population mean hours of TV watching among Muslims is likely to be between 0.3 and 4.3.
c)   The confidence interval is wide mainly because of the very small sample size.

⌨**8.35**   **Grandmas using e-mail:**
a)   Mean = 4.14; standard deviation =5.08; standard error=$5.08/\sqrt{7} = 1.92$.
b) $4.14 \pm 1.943(1.92)$, which is 0.4 to 7.9. We are 90% confident that the population mean number of hours per week spent sending and answering e-mail for women of at least age 80 is between 0.4 and 7.9 hours.
c)   Since there will be many women of age 80 or older who do not use email at all but some who use email a lot, the distribution is likely to be skewed right. Since the confidence interval using the t-distribution is a robust method, the interval in part b) should still be valid.

8.37   **More eBay selling prices:**
The most plausible values for the population mean with the buy-it-now option are between \$231.91 and \$238.99. The most plausible values for the population mean without the buy-it-now option are between \$219.32 and \$227.40. It seems that the buy-it-now option leads to higher prices.

8.39   **Political views:**
a)   To construct the confidence interval, we would first calculate the margin of error. The margin of error equals the standard error, 0.0215, multiplied by the *t*-score for 4332 degrees of freedom ($n − 1$) and a 95% confidence interval (1.96). To calculate the confidence interval, we would then subtract the margin of error from the mean, 4.12, to get the lower limit of the confidence interval and add it to the mean to get the higher limit of the confidence interval.
b)   We can conclude that the population mean is higher than 4.0 because 4.0 falls below the lower endpoint of the confidence interval.
c)   (i)   A 99% confidence interval would be wider than a 95% confidence interval.
       (ii)   A smaller sample size would lead to a wider confidence interval than would a larger sample size.

8.41   **Effect of n:**

The margin of error is calculated by multiplying the standard error by the appropriate t-score. To get the standard error, we must divide the standard deviation by the square root of n.

a) $se = 100/\sqrt{400} = 5$; margin of error $= (1.966)(5) = 9.8$

b) $se = 100/\sqrt{1600} = 2.5$; margin of error $= (1.961)(2.5) = 4.9$

As the sample size increases, the margin of error becomes smaller.

8.43   **Catalog mail-order sales:**

a)   It is not plausible that the population distribution is normal because a large proportion are at the single value of 0. Because we are dealing with a sampling distribution of a sample greater than size 30, this is not likely to affect the validity of a confidence interval for the mean. Large random samples lead to sampling distributions of the sample mean that are approximately normal.

b)   The mean is $10, $se = 10/\sqrt{100} = 1$, margin of error $= (1.984)(1) = 1.98$

Confidence interval: (8.0, 12.0)

It does seem that the sales per catalog declined with this issue. $15 is not in the confidence interval, and, therefore, is not a plausible population mean for the population from which this sample came.

🖥**8.45**   **Simulating the confidence interval:**

a)   The results will differ each time this exercise is conducted.

b)   We would expect 5% of the intervals not to contain the true value.

c)   Close to 95% of the intervals contain the population mean even though the population distribution is quite skewed. This is so because with a large random sample size, the sampling distribution is approximately normal even when the population distribution is not. The assumption of a normal population distribution becomes less important as n gets larger.

SECTION 8.4: PRACTICING THE BASICS

8.47   **Binge drinkers:**

$$n = \frac{\hat{p}(1-\hat{p})z^2}{m^2} = \frac{0.44(1-0.44)1.96^2}{0.05^2} = 379$$

8.49   **How many businesses fail?:**

a) $n = \dfrac{\hat{p}(1-\hat{p})z^2}{m^2} = \dfrac{0.5(1-0.5)1.96^2}{0.10^2} = 96.$

b) $n = \dfrac{\hat{p}(1-\hat{p})z^2}{m^2} = \dfrac{0.5(1-0.5)1.96^2}{0.05^2} = 384.$

c) $n = \dfrac{\hat{p}(1-\hat{p})z^2}{m^2} = \dfrac{0.5(1-0.5)2.58^2}{0.05^2} = 666.$

d)   As the margin of error decreases, we need a larger sample size to guarantee estimating the population proportion correct within the given margin of error and at a given confidence level. As the confidence level increases, we also need a larger sample size to get the desired results.

8.51   **Farm size:**

a) $n = \dfrac{\sigma^2 z^2}{m^2} = \dfrac{(200)^2(1.96)^2}{25^2} = 246.$

b)   We can use the same formula as in part (a). $246 = \dfrac{(300)^2(1.96)^2}{m^2}$

$m^2 = (1.96)^2(300)^2/(246) = 1405.46$

$m = 37.5$

8.53 **Population variability**:
For a very diverse population, we'd have a wider range of observed values, and hence, a larger standard deviation. Larger standard deviations result in larger standard errors and wider confidence intervals. For a homogenous population, we would have a smaller standard deviation, and would not need the large sample size for the denominator of the standard error formula. When estimating the mean income for all medical doctors in the U.S., we'd have a fairly wide income: from the lower range of incomes of rural family doctors to the extremely high range of incomes of specialists at major teaching hospitals. A sample from this population would likely have a large standard deviation. For a population of McDonald's entry- level employees in the U.S., however, we'd have a much smaller range. They'd all likely be making minimum wage or slightly higher. The standard deviation of a sample from this population would be relatively small.

8.55 **Do you like tofu?**
a) The sample proportion is 5/5 = 1.0.

b) $se = \sqrt{\hat{p}(1-\hat{p})/n} = \sqrt{1(1-1)/5} = 0$; the usual interpretation of standard error does not make sense. The sampling distribution is likely, after all, to have some variability because the true probability (which determines the exact standard error) is positive.

c) The margin of error is (1.96)(0) = 0, and thus, the confidence interval is 1.0 to 1.0. It is not sensible to conclude that all students at the school like tofu because this method works poorly in this case.

d) It is not appropriate to use the large-scale confidence interval in (c) because we do not have more than 15 successes and 15 failures. It is more appropriate to add two to the successes and the failures and then repeat the process. This would give seven who said they liked it, and two who said they did not, for a total of 9. The new sample proportion would be 7/9 = 0.78.

$$se = \sqrt{\hat{p}(1-\hat{p})/n} = \sqrt{0.78(1-0.78)/9} = 0.138$$

margin of error = (1.96)(0.138) = 0.270

confidence interval: 0.51 to 1.05 (or 1.0 because p cannot exceed 1)

We can be 95% confident that the proportion of students who like tofu is within this interval.

8.57 **Accept a credit card?**:
First, add two to the successes and failures. The sample proportion is 2/104 = 0.019.

$$se = \sqrt{\hat{p}(1-\hat{p})/n} = \sqrt{0.019(1-0.019)/104} = 0.013$$

Margin of error = (1.96)(0.013) = 0.025

Confidence interval: - 0.006 to 0.044 (Actually, the lower limit would be zero because we cannot have a proportion below zero; round to 0.00 and 0.04.)

They can conclude that fewer than 10% of their population would take the credit card.

SECTION 8.5: PRACTICING THE BASICS

8.59 **Estimating variability**:
We would sample with replacement from the 10 weight values, taking 10 observations and finding the standard deviation. We would do this many, many times – 10,000 times, perhaps. The 95% confidence interval would be the values in the middle 95% of the standard deviation values (from 2.5% to 97.5%).

CHAPTER PROBLEMS: PRACTICING THE BASICS

8.61 **Divorce and age of marriage:**
a) These are point estimates.
b) The information here is not sufficient to construct confidence intervals. We also need to know sample sizes.

8.63   **British monarchy:**

The first mean sample proportion is 0.21. It has a standard error of $se = \sqrt{\hat{p}(1-\hat{p})/n} =$

$\sqrt{0.21(1-0.21)/3000} = 0.0074$; thus, the confidence interval is $\hat{p} \pm z(se)$, with a z of 1.96 for the 95% confidence interval. 0.21 - (1.96)(0.0074) = 0.195; 0.21 + (1.96)(0.0074) = 0.225. The margin of error for each of these is 0.015. The confidence interval is (0.195, 0.225)

The second mean sample proportion is 0.53. It has a standard error of $se = \sqrt{\hat{p}(1-\hat{p})/n} =$

$\sqrt{0.53(1-0.53)/3000} = 0.0091$; thus, the confidence interval is $\hat{p} \pm z(se)$, with a z of 1.96 for the 95% confidence interval. 0.53 - (1.96)(0.0091) = 0.512; 0.53 + (1.96)(0.0091) = 0.548. The margin of error for each of these is 0.018. The confidence interval is (0.512, 0.548)

8.65   **Life after death:**

We could form the confidence interval by subtracting the 1.4% margin of error from 82.8% to get the lower limit, and adding the 1.4% margin of error to 82.8% get the upper limit. We can say that we are 95% confident that the population percentage of people who believe in life after death falls between 81.4 and 84.2%.

8.67   **Vegetarianism:**
a) We must assume that the data were obtained randomly.

b) $se = \sqrt{\hat{p}(1-\hat{p})/n} = \sqrt{0.04(1-0.04)/10,000} = 0.002$

    The confidence interval is $\hat{p} \pm z(se)$

    Lower limit: 0.04 − 2.58(0.002) = 0.035

    Upper limit: 0.04 + 2.58(0.002) = 0.045

    The interval is so narrow, even though the confidence level is high, mainly because of the very large sample size. The very large sample size contributes to a small standard error by providing a very large denominator for the standard error calculation.

c) We can conclude that fewer than 10% of Americans are vegetarians because 10% falls above the highest believable value in the confidence interval.

8.69   **Population data:**

It doesn't make sense to construct a confidence interval because we have data for the entire population. We can actually <u>know</u> the proportion of vetoed bills; we don't have to estimate it.

8.71   **Legalize marijuana?**
a) 672 said "legal." 1,156 said "not legal." The sample proportions are 0.3676 and 0.6324, respectively.

b) $se = \sqrt{\hat{p}(1-\hat{p})/n} = \sqrt{0.3676(1-0.3676)/1828} = 0.0113$.

    A 95% confidence interval is $\hat{p} \pm z(se)$

    Lower limit: 0.3676 − 1.96(0.0113) = 0.35

    Upper limit: 0.3676 + 1.96(0.0113) = 0.39

    We can conclude that a minority of the population supports legalization because 0.50 is above the upper endpoint of the 95% confidence interval.

c) It appears that the proportion favoring legalization is increasing over time.

8.73   **Nondrinkers:**

The first column, "Sample," refers to the specific sample under study: in this case, non drinkers, abbreviated as "Nodrink." The next column, "X," is the number in the sample who said that they have been lifetime abstainers from drinking alcohol, 7380 people. The third column, "N," is the sample size of 30,000 people. The "Sample p" is the proportion of the total sample, 0.246, who said that they had been lifetime abstainers from drinking alcohol. The final column, "95.0% CI," is the 95% confidence interval for this proportion. We can be 95% confident that the population proportion falls between 0.241 and 0.251.

8.75   Grandpas using e-mail:

a) The first result is a 90% confidence interval for the mean hours spent per week sending and answering e-mail for males of at least age 75. The sample mean, $\bar{x}$, is listed as 6.38 hours. Thus, the estimated mean hours spent per week sending and answering e-mail for males of at least age 75 is 6.38 hours. The sample standard deviation is 6.02. This quantity estimates the population standard deviation which tells us how far we can expect a typical observation to vary from the mean. These estimates are based on a sample of size 8.

b) The confidence interval is given as 2.34 to 10.41. We can be 90% confident that the population mean number of hours spent per week sending and answering email for males of at least age 75 is between 2.34 and 10.41 hours.

c) Since there are likely to be a lot of men over the age of 75 who do not use e-mail but also some who use e-mail regularly, this distribution is likely skewed right. Since the *t*-distribution is robust to violations of the normality assumption, the interval is still valid.

8.77   *t*-scores:

a) The *t*-scores for a 95% confidence interval for a sample size of 10 is 2.262, for a sample size of 20 is 2.093, for a sample size of 30 is 2.045, and for a sample size of infinity is 1.96 (the *z*-score for a 95% confidence interval).

b) The answer in (a) suggests that the *t* distribution approaches the standard normal distribution as the sample size gets larger.

8.79   Psychologists' income:

a) $se = s/\sqrt{n} = 16870/\sqrt{31} = 3{,}029.94$. Confidence interval $\bar{x} \pm t(se)$ has bounds 43,834 – (2.042)(3,029.94) = 37,647 and 43,834 + (2.042)(3,029.94) = 50,021
We can be 95% confident that the population mean income is between $37,647 and $50,021.

b) It assumes an approximately normal population distribution.

c) If the assumption about the shape of the population distribution is not valid, even with a small *n* the results aren't necessarily invalidated. The method is robust in terms of the normal distribution assumption. However, if there were some extreme outliers, this might not hold true.

8.81   How long lived in town?:

a) The population distribution is not likely normal because the standard deviation is almost as large as the mean. In fact, the lowest possible value of 0 is only 20.3/18.2 = 1.1 standard deviations below the mean. Moreover, the mean is quite a bit larger than the median. Both of these are indicators of skew to the right.

b) We can construct a 95% confidence interval, however, because the normal population assumption is much less important with such a large sample size.

$se = 18.2/\sqrt{1415} = 0.484$. Confidence interval $\bar{x} \pm t(se)$ has bounds 20.3 – (1.962)(0.484) = 19.4 and 20.3 + (1.962)(0.484) = 21.2.
We can be 95% confident that the population mean number of years lived in a given city, town, or community is between 19.4 and 21.2.

8.83   How often feel sad?:

The mean is 1.4, and the standard deviation is as follows: $s = \sqrt{\dfrac{\sum(x-\bar{x})^2}{n-1}} =$

$$\sqrt{\frac{(0-1.4)^2+(0-1.4)^2+(1-1.4)^2+(0-1.4)^2+(7-1.4)^2+(2-1.4)^2+(1-1.4)^2+(0-1.4)^2+(0-1.4)^2+(3-1.4)^2}{10-1}}$$

$= 2.221$; The standard error: $se = s/\sqrt{n} = 2.221/\sqrt{10} = 0.702$

90% confidence interval: $\bar{x} \pm t(se)$ has endpoints 1.4 – (1.833)(0.702) = 0.1 and 1.4 + (1.833)(0.702) = 2.7. We can be 95% confident that the mean for the population of Wisconsin students is between 0.1 and 2.7. For this inference to apply to the population of all University of Wisconsin students, we must assume that the data are randomly produced and that we have an approximately normal population distribution.

⌨8.85 Revisiting mountain bikes:
a) We can be 95% confident that the population mean for mountain bike price falls between $411 and $845.
b) To form the interval, we need to assume that the data are produced randomly, and that the population distribution is approximately normal. The population does not seem to be distributed normally. The data seem to cluster on the low and high ends, with fewer in the middle. Unless there are extreme outliers, however, this probably does not have much of an effect on this inference. This method is fairly robust with respect to the normal distribution assumption.

⌨8.87 Income for families in public housing:
a)

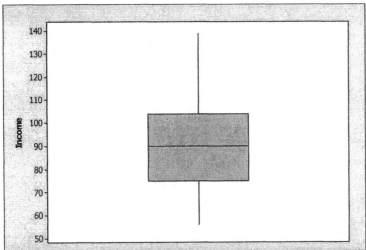

From the box plot, we can predict that the population distribution is skewed to the right. There does not seem to be an extreme outlier, so this should not affect the population inferences. This method is fairly robust with respect to the normal distribution assumption.
b) According to software, the mean is 90.2 and the standard deviation is 20.1.
c) We can be 95% confident that the mean income for the population of families living in public housing in Chicago is between $8,256 and $9,822.

8.89 Males watching TV:
We can be 95% confident that the mean number of hours spent watching TV for the population of males falls between 2.7 and 3.0.

8.91 Highest grade completed:
We could make inferences using the mean of 13.3 (standard deviation of 3.2), or using the proportion of 0.17. For the proportion,

$$se = \sqrt{\hat{p}(1-\hat{p})/n} = \sqrt{0.17(1-0.17)/4499} = 0.0056.$$

Confidence interval: $\hat{p} \pm z(se)$ has bounds $0.17 - (1.96)(0.0056) = 0.159$; $0.17 + (1.96)(0.0056) = 0.181$, giving the confidence interval: (0.16, 0.18).

8.93 Sex partners in previous year:
a) $se = s/\sqrt{n} = 1.09/\sqrt{2400} = 0.022.$
b) The distribution was probably highly skewed to the right because the standard deviation is larger than the mean. In fact, the lowest possible value of 0 is only $1.04/1.09 = 0.95$ standard deviation below the mean.
c) The skew need not cause a problem with constructing a confidence interval for the population mean, unless there are extreme outliers, because this method is robust with respect to the normal distribution assumption.

8.95 **Driving after drinking**:

a) $n = [\,\hat{p}\,(1-\hat{p}\,)z^2\,]/m^2 = [0.2(1\text{-}0.2)(1.96)^2]/(0.04)^2 = 384$

b) $n = [\,\hat{p}\,(1-\hat{p}\,)z^2\,]/m^2 = [0.5(1\text{-}0.5)(1.96)^2]/(0.04)^2 = 600$

This is larger than the answer in (a). If we can make an educated guess about what to expect for the proportion, we can use a smaller sample size, saving possibly unnecessary time and money.

8.97 **Mean property tax**:

a) $n = \dfrac{\sigma^2 z^2}{m^2} = \dfrac{(1000)^2 (1.96)^2}{100^2} = 385.$

The solution makes the assumption that the standard deviation will be similar now.

b) The margin of error would be more than $100 because the standard error will be larger than predicted.

c) With a larger margin of error, the 95% confidence interval is wider; thus, the probability that the sample mean is within $100 (which is less than the margin of error from b)) of the population mean is less than 0.95.

CHAPTER PROBLEMS: CONCEPTS AND INVESTIGATIONS

8.99 **Religious beliefs**:

Each student's one-page report will be different, but will explain the logic behind random sampling and the effect of sample size on margin of error.

8.101 **Housework and gender**:

Men: $se = s/\sqrt{n} = 12.9/\sqrt{4252} = 0.198$. 95% confidence interval has endpoints $18.1 - (1.96)(0.198) = 17.7$ and $18.1 + (1.96)(0.198) = 18.5$.

Women: $se = s/\sqrt{n} = 18.2/\sqrt{6764} = 0.221$. 95% confidence interval has endpoints $32.6 - (1.96)(0.221) = 32.2$ and $32.6 + (1.96)(0.221) = 33.0$.

We can compare men and women by creating a confidence interval for each gender. The assumptions on which the confidence interval is based are that the data were randomly produced and that the population distributions are approximately normal. It seems that women do more housework than do men, on the average.

8.103 **Types of estimates**:

If we know the confidence interval of (4.0, 5.6), we know the mean falls in the middle because the confidence interval is calculated by adding and subtracting the same number from the mean. In this case, the mean equals 4.8. On the other hand, if we only knew the mean of 4.8, we could not know the confidence interval, and would have much less of an idea of how accurate this point estimate is likely to be.

8.105 **99.9999% confidence**:

An extremely large confidence level makes the confidence interval so wide as to have little use.

⌨**8.107** **Outliers and CI**:

When the observation of 6 is changed to 60, the 95% confidence interval is now (-10.4, 30.4). This is a much wider interval; an outlier can dramatically affect the standard deviation and the standard error.

8.109 **CI property**:

The best answer is (a).

8.111 **Number of close friends**:

Both (b) and (e) are correct.

8.113   Mean age at marriage:
a)   The confidence interval refers to the population, not the sample, mean.
b)   The confidence interval is an interval containing possible means, not possible individual scores.
c) $\bar{x}$ is the sample mean; we know exactly what it is.
d)   If we sampled the entire population even once, we would know the population mean exactly.

8.115   True or false 1:
False. It should be the population proportion.

8.117   True or false 3:
False. A volunteer sample is not a random sample, thus violating one of the necessary assumptions.

♦♦8.119 Opinions over time about the death penalty:
a)   When we say we have 95% confidence in an interval for a particular year, we mean that in the long-run (that is, if we took many random samples of this size from this population), the intervals based on these samples would capture the true population proportion 95% of the time.
b)   The probability of all intervals containing the population mean is

$$p(x) = \frac{n!}{x!(n-x)!}p^x(1\text{-}p)^{n-x} = \frac{20!}{20!(20-20)!}\,0.95^{20}(1\text{-}0.95)^{20-20} = 1(0.358)(1) = 0.358$$

c)   The mean of the probability distribution of X is $(n)(p) = (20)(0.95) = 19$
d)   To make it more likely that all 20 inferences are correct, we could increase the confidence level, to 0.99 for example.

♦♦8.121 Estimating p without estimating *se*:
a) $|\hat{p} - p| = 1.96\sqrt{p(1-p)/n}$

$|0 - p| = 1.96\sqrt{p(1-p)/30}$

If we substitute 0, we get 0 on both sides of the equation.
If we substitute 0.1135, we get 0.1135 (the absolute value of $-$ 0.1135) on the left, and on the right.
b)   The confidence interval formed using the method in this exercise seems more believable because it forms an actual interval and contains more than just the value of 0. It seems implausible that there are no vegetarians in the population.

♦♦8.123 Median as point estimate:
If the population is normal, the standard error of the median is 1.25 times the standard error of the mean. A larger standard error means a larger margin of error, and therefore, a wider confidence interval. The sample mean tends to be a better estimate than the sample median because it is more precise.

CHAPTER PROBLEMS: STUDENT ACTIVITIES

▣8.125 GSS project:
The results will be different each time this exercise is conducted.

Chapter 9
Statistical Inference: Significance Tests about Hypotheses

SECTION 9.1: PRACTICING THE BASICS

9.1 H_0 or H_a ?:
  a) null hypothesis
  b) alternative hypothesis
  c) (a) $H_0 : p = 0.50$; $H_a : p \neq 0.50$
      (b) $H_0 : p = 0.24$; $H_a : p < 0.24$

9.3   **Burden of proof:**

H_0 : The pesticide is not harmful to the environment.

H_a : The pesticide is harmful to the environment.

9.5   **Low Carbohydrate Diet:**
  a) This is an alternative hypothesis because it has a range of parameter values.
  b) The relevant parameter is the mean weight change, μ. H_0 : $\mu = 0$; this is a null hypothesis.

9.7   **z test statistic:**
The data give strong evidence against the null hypothesis. Most scores fall within three standard errors of the mean, and this sample proportion falls over three standard errors from the null hypothesis value.

SECTION 9.2: PRACTICING THE BASICS

9.9   **Psychic:**
The null hypothesis is that the psychic will predict the outcome of the roll of a die in another room 1/6 of the time, and the alternative hypothesis is that the psychic will predict the outcome more than 1/6 of the time.
$H_0 : p = 1/6$ and $H_a : p > 1/6$

9.11   **Get P-value from z:** (Note: Use Table A)
  a) 0.15
  b) 0.30
  d) None of these P-values gives strong evidence against H_0. All of them indicate that the null hypothesis is plausible.

9.13   **Find test statistic and P-value:**
  a) standard error = $\sqrt{p_0(1-p_0)/n} = \sqrt{0.5(1-0.5)/100} = 0.05$
      $z = (0.35 - 0.50)/0.05 = -3.0$
  b) The P-value is 0.001.
  c) If the null hypothesis were true, the probability would be 0.001 of getting a test statistic at least as extreme as the value observed. This does provide strong evidence against H_0. This is a very small P-value; it does not seem plausible that $p = 0.50$.

9.15 **Religion important in your life?:**
1. The response is categorical with outcomes "yes" or "no" to the statement that religion is at least "somewhat important" in your life; p represents the probability of a yes response. The poll was a random sample of 2546 18-24-year-olds and $np_0 = n(1-p_0) = 2546(0.5) = 1273 \geq 15$.
2. $H_0: p = 0.5$; $H_a: p > 0.5$. To determine whether a majority of 18-24-year-olds believe that religion is at least "somewhat important" in their lives, we will test the alternative hypothesis that this probability is greater than 0.5.
3. $\hat{p} = 0.70$ so $z = \dfrac{\hat{p} - p_0}{\sqrt{p_0(1-p_0)/n}} = \dfrac{0.70 - 0.50}{\sqrt{0.5(0.5)/2546}} = 20.2$. The sample proportion is 20.2 standard errors above the null hypothesis value.
4. The P-value ≈ 0 is the probability of obtaining a sample proportion at least as extreme as the one observed, if the null hypothesis is true.
5. Since the P-value is approximately 0, the sample supports the alternative hypothesis. There is very strong evidence that a majority of 18-24-year-olds believe that religion is at least "somewhat important" in their lives.

9.17 **Another test of therapeutic touch:**
a) p = proportion of trials guessed correctly
   $H_0: p = 0.50$ and $H_a: p > 0.50$
b) $p = 53/130 = 0.4077$
   $se = \sqrt{p_0(1-p_0)/n} = \sqrt{0.50(1-0.50)/130} = 0.0439$
   $z = (0.4077 - 0.5)/0.0439 = -2.10$; the sample proportion is a bit more than 2 standard errors less than would be expected if the null hypothesis were true.
c) The P-value is 0.98. Because the P-value is larger than the significance level of 0.05, we do not reject the null hypothesis. The probability would be 0.98 of getting a test statistic at least as extreme as the value observed if the null hypothesis were true, and the population proportion were 0.50.
d) $np = (130)(0.5) = 65 = n(1-p)$; the sample size was large enough to make the inference in (c). We also would need to assume randomly selected practitioners and subjects for this to apply to all practitioners and subjects.

9.19 **Gender bias in selecting managers:**
a) $H_0: p = 0.60$ and $H_a: p > 0.60$ where p is the probability of selecting a male.
   The company's claim is a lack of gender bias is a "no effect" hypothesis because they predict that the proportion of males chosen for management training is the same as in the eligible pool.
b) The large-sample analysis is justified because the expected successes and failures are both at least fifteen under H_0; $np = (40)(0.6) = 24$, and $n(1-p) = (40)(0.4) = 16$. The software obtained the test-statistic value by subtracting the p predicted by the null hypothesis from the p observed in the sample, then dividing that by the standard error.
c) The P-value in the table refers to the alternative hypothesis that $p \neq 0.60$. For the one-sided alternative, it is half this, or about 0.10.
d) For a 0.05 significance level, we would not reject the null hypothesis because the P-value > 0.05. It is plausible that the null hypothesis is correct, and that there is not a greater proportion of male trainees than would be expected based on the eligible pool.

9.21 **Garlic to repel ticks:**
a) The relevant variable is whether garlic or placebo is more effective, and the parameter is the population proportion, p = those for whom garlic is more effective than placebo.
b) $H_0: p = 0.50$ and $H_a: p \neq 0.50$; the sample size is adequate because there are at least 15 successes (garlic more effective) and failures (placebo more effective).

c) $\hat{p} = 37/66 = 0.561$

$se = \sqrt{p_0(1-p_0)/n} = \sqrt{0.50(1-0.50)/66} = 0.062$

$z = (0.561 - 0.5)/0.062 = 0.984$

d) The P-value is 0.33. This P-value is not that extreme. If the null hypothesis were true, the probability would be 0.33 of getting a test statistic at least as extreme as the value observed. It is plausible that the null hypothesis is correct. Because the probability is 0.33 that we would observe our test statistic or one more extreme due to random variation, there is not strong evidence that the population proportion which would have fewer tick bites with garlic vs. placebo differs from 0.50.

9.23 Which cola?:

a) The test statistic ("Z-Value") is calculated by taking the difference between the sample proportion and the null proportion and dividing it by the standard error.

b) We get the "P-value" by looking up the "Z-value" in Table A or using software. We have to determine the two-tail probability from the standard normal distribution below -1.286 and above 1.286. The P-value of 0.20 tells us that if the null hypothesis were true, a proportion of 0.20 of samples would fall at least this far from the null hypothesis proportion of 0.50. This is not very extreme; it is plausible that the null hypothesis is correct, and that Coke is not preferred to Pepsi.

c) It does not make sense to accept the null hypothesis. It is possible that there is a real difference in the population that we are not detecting in our test (perhaps because the sample size is not very large), and we can never accept a null hypothesis. A confidence interval shows that 0.50 is one of many plausible values.

d) The 95% confidence interval tells us the range of plausible values, whereas the test merely tells us that 0.50 is plausible.

♦♦9.25 A binomial headache:

This P-value gives strong evidence against the null hypothesis. It would be very unlikely to have a sample proportion of 1.00 if the actual population proportion were 0.50.

SECTION 9.3: PRACTICING THE BASICS

9.27 Which *t* has P-value = 0.05?:

a) 2.145

b) 1.762

c) - 1.762

9.29 Effect of *n*:

a) The P-value would be larger when $t = 1.20$ than when $t = 2.40$ because the *t*-value of 1.20 is less extreme.

b) Larger sample sizes result in smaller P-values because they decrease the standard error (by making its denominator larger). Standard error becomes the denominator when calculating the test statistic. A smaller denominator leads to a larger test statistic which leads to smaller P-values.

9.31 Men at work:

a) The relevant variable is the number of hours worked in the previous week by male workers; the parameter of interest is the population mean work week (in hours) for men.

b) H_0: $\mu=40$; H_a: $\mu>40$.

c) P-value ≈ 0. The P-value is the probability of obtaining a sample with a mean of 45.3 or more hours assuming the null hypothesis were true.

d) Since the P-value is less than the significance level of 0.01, there is sufficient evidence to reject the null hypothesis and to conclude that the population mean work week for men exceeds 40 hours.

9.33 Lake pollution:

a) From software, the mean of the four observations is 2000; the standard deviation is 816.5; and the standard error is 408.25.

b) Software gives a *t*-score of 2.45.

c) The P-value is 0.046 for a one-sided test. This is smaller than 0.05, so we have enough evidence to reject the null hypothesis at a significance level of 0.05. We have relatively strong evidence that the wastewater limit is being exceeded.

d) The one-sided analysis in (b) implicitly tests the broader null hypothesis that $\mu \leq 1000$. We know this because if it would be unusual to get a sample mean of 2000 if the population mean were 1000, we know that it would be even more unusual to get this sample mean if the population mean were less than 1000.

9.35   Crossover study:

a) The difference scores are 40, 15, 90, 50, 30, 70, 20, 30, -35, 40, 30, 80, 130. The sample size is small so it we cannot tell too much from the plot, but there is no evidence of severe non-normality

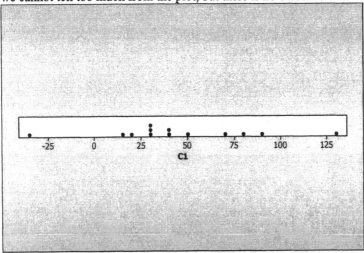

b) 1) Assumptions: The data (PEF change scores) are randomly obtained from a normal population distribution. Here, the data are not likely produced using randomization, but are likely a convenience sample. The two-sided test is robust if the population distribution is not normal

2) Hypotheses: $H_0: \mu = 0; H_a: \mu \neq 0$

3) Test statistic (mean and standard deviation of sample calculated using software):

$$t = \frac{(45.4 - 0)}{40.6 / \sqrt{13}} = 4.03$$

4) P-value: 0.002

5) Conclusion: If the null hypothesis were true, the probability would be 0.002 of getting a test statistic at least as extreme as the value observed. There is strong evidence that PEF levels were lower with salbutamol than with formoterol.

c) The assumption of random production of does not seem valid for this example. A convenience sample limits reliability in applying this inference to the population at large.

9.37   Selling a burger:

1) Assumptions: The data are produced using randomization, from a normal population distribution. The two-sided test is robust for the normality assumption.

2) Hypotheses: $H_0: \mu = 0; H_a: \mu \neq 0$

3) Test statistic: $t = \dfrac{(3000 - 0)}{4000 / \sqrt{10}} = 2.37$

4) P-value: 0.04

5) Conclusion: If the null hypothesis were true, the probability would be 0.04 of getting a test statistic at least as extreme as the value observed. Because the P-value of 0.04 is < 0.05, there is sufficient evidence that the coupons led to higher sales than did the outside posters.

⌨9.39 **Anorexia in teenage girls**:

a) Most of the data fall between 4 and 14. The sample size is small so it we cannot tell too much from the plot, but there is no evidence of severe non-normality

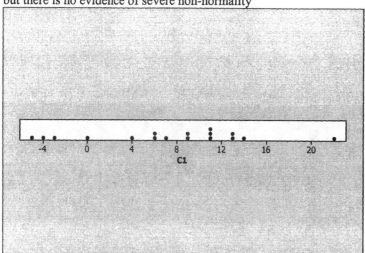

b) Software verifies these statistics.

c) 1) The necessary assumptions are that the data are quantitative and are produced randomly and the population distribution should be approximately normal.

2) $H_0: \mu = 0; H_a: \mu \neq 0$

3) The test statistic, according to software, is 4.19.

4) The P-value is 0.001.

5) This extreme P-value suggests that we have strong evidence against the null hypothesis that family therapy has no effect. If the null hypothesis were true, the probability would be only 0.001 of getting a test statistic at least as extreme as the value observed.

9.41 **Test and CI**:

Results of 99% confidence intervals are consistent with results of two-sided tests with significance levels of 0.01. A confidence interval includes the most plausible values for the population mean to a 99% degree of confidence. If the test rejects the null hypothesis with significance level 0.01, then the 99% confidence interval does not contain the value in the null hypothesis.

SECTION 9.4: PRACTICING THE BASICS

9.43 **Error probability**:

a) The probability of Type I error would be 0.05.

b) If this test resulted in a decision error, it was a Type I error.

9.45 **Anorexia errors**:

a) A Type I error would occur if we rejected the null hypothesis when it was true. Thus, we concluded that the therapy had an effect when in fact it did not.

b) A Type II error would occur if we failed to reject the null hypothesis when it was false. Thus, we determined that it was plausible that the null hypothesis was correct and that the therapy might have no effect, when in fact it does.

9.47 **Errors in the courtroom**:

a) If H_0 is rejected, we conclude that the defendant is guilty.

b) A Type I error would result in finding the defendant guilty when he/she is actually innocent.

c) If we fail to reject H_0, the defendant is found not guilty.

d) A Type II error would result in failing to convict a defendant who is actually guilty.

9.49 **Decision errors in medical diagnostic testing:**
a) A Type I error is a false positive because we have rejected the null hypothesis that there is no disease, but we were wrong. The woman in fact does not have breast cancer. The consequence would be that the woman would have treatment, or at least further testing, when she did not need any.
b) A Type II error is a false negative because we have failed to reject the null hypothesis that there is no disease, but we were wrong. The woman in fact does have breast cancer. The consequence would be failing to detect cancer and treat the cancer when it actually exists.
c) The disadvantage of this tactic is that more women who <u>do</u> have breast cancer will have false negative tests and not receive necessary treatment.

9.51 **Which error is worse?:**
a) When rejecting the null results in the death penalty, a Type I error is worse than a Type II error. With a Type II error, a guilty man or woman goes free, whereas with a Type I error, an innocent man or woman is put to death.
b) When rejecting the null hypothesis results in treatment for breast cancer, a Type II error is worse than a Type I error. With a Type I error, someone might receive additional tests (e.g., biopsy) before ruling out breast cancer, but with a Type II error, someone might not receive life-saving treatment when they need it.

SECTION 9.5: PRACTICING THE BASICS

9.53 **Practical significance:**
a) Test statistic: $t = \dfrac{(498 - 500)}{100 / \sqrt{40,000}} = -4.0$.
b) P-value $= 2P(Z < -4.0) = 0.00006$.
c) This result is statistically significant because the P-value is very small, but it is not practically significant because the sample mean of 498 is very close to the null hypothesis mean of 500.

9.55 **Fishing for significance:**
This is misleading because, with a significance level of 0.05, we would expect 5% of tests to be significant just by chance if the null hypothesis is true, and for 60 tests this is $0.05(60) = 3$ tests.

9.57 **How many medical discoveries are Type I errors?:**
The following tree diagram is based on 100 studies.

                     True effect?   **Decision: Reject H₀?**

                                             Yes (14)

              Yes (20) -------

                                             No (6)

 100 ------------

                                           Yes (4)

              No (80) --------

                                           No (76)

The proportion of actual Type I errors (of cases where the null is rejected) would be about $4/(4 + 14) = 0.22$.

SECTION 9.6: PRACTICING THE BASICS

9.59 **Two sampling distributions:**
a) A one-tailed test would have a z-score of 1.645 at the cutoff. Here, the standard error would be $\sqrt{p_0(1 - p_0)/n} = \sqrt{0.5(1 - 0.5)/100} = 0.050$. The value 1.645 standard errors above 0.50 is $0.50 + 1.645(0.050) = 0.582$.
c) $z = \dfrac{(0.582 - 0.60)}{\sqrt{.60(1 - .60)/100}} = -0.37$; Table A tells us that 0.36 falls beyond this z-score; thus, 0.36 is the probability of a Type II error.

9.61 **Balancing Type I and Type II errors**:

a) The cutoff for a 0.01 significance level and a one-tailed test is 2.33. Here, the standard error would be $\sqrt{p_0(1-p_0)/n} = \sqrt{0.333(1-0.333)/116} = 0.0438$. The value 2.33 standard errors above 0.333 is 0.333 + 2.33(0.0438) = 0.435.

b) If $p = 0.50$, the z-score for 0.435 in reference to 0.50 is $z = \dfrac{(0.435 - 0.5)}{\sqrt{0.5(1-0.5)/116}} = -1.40$. If we look this z-score up on a table, we find that the proportion of this curve that is not in the rejection area is 0.08.

9.63 **Type II error with two-sided H$_a$:**

a) The standard error is $\sqrt{\dfrac{p_0(1-p_0)}{n}} = 0.0438$, where $p_0 = 1/3$.

For H$_a$, a test statistic of $z = 1.96$ has a P-value (two-tail probability) of 0.05. We reject H$_0$ when $|\hat{p} - 1/3| \geq 1.96(se) = 0.086$, hence we need
$\hat{p} \geq 0.086 + 1/3 = 0.419$ or $\hat{p} \leq 1/3 - 0.086 = 0.248$

When H$_0$ is false, a Type II error occurs if $0.248 < \hat{p} < 0.419$.

b) We can calculate z-scores for each of these proportions. $z = (0.248 - 0.50)/0.0464 = -5.43$; the probability that $\hat{p}$ is less than this z-score is 0. $z = (0.419 - 0.50)/0.0464 = -1.75$; the probability that $\hat{p}$ is greater than this z-score is 0.96.

c) The probability of a Type II error is the portion of the curve (for the parameter 0.50) that is not over the rejection area. This is 0.04.

🖥9.65 **Simulating Type II errors**:

a) Results will be different each time this exercise is conducted. Theory predicts that the P(Type II error) will be larger with a p that is closer to that predicted by the null hypothesis. In other words, it will be larger than 0.02.

b) The proportion of Type II errors will be smaller with $p = 0.50$ than with $p = 0.45$. The results of the simulation will be different each time this exercise is conducted.

c) If the sample size is smaller, the standard error will be larger and the proportions at the cutoffs will be further from the proportion in H$_0$ .This will increase the probability of a Type II error.

CHAPTER PROBLEMS: PRACTICING THE BASICS

9.67 **ESP**:

1) Assumptions: The data are categorical (correct vs. incorrect guesses) and are obtained randomly. The expected successes and failures are less than fifteen under H$_0$; $np = (20)(0.5) = 10 < 15$, and $n(1-p) = (20)(0.5) = 10 < 15$, so this test is approximate.

2) Hypotheses: H$_0$: $p = 0.50$; H$_a$: $p > 0.50$

3) Test statistic: $z = \dfrac{0.60 - 0.50}{\sqrt{0.5(1-0.5)/20}} = 0.89$

4) P-value: 0.19.

5) Conclusion: If the null hypothesis were true, the probability would be 0.19 of getting a test statistic at least as extreme as the value observed. There is not strong evidence with a P-value of 0.19 that the population proportion of correct guesses is higher than 0.50.

9.69 Box or Draper?:

a) 1) Assumptions: The data are categorical (Box and Draper) and are obtained randomly; the expected successes and failures are both at least fifteen under H_0; $np = (0.5)(400) \geq 15$, and $n(1-p) = (0.5)(400) \geq 15$.

2) p = population proportion of voters who prefer Box; Hypotheses: H_0: $p = 0.50$; H_a : $p \neq 0.50$

3) Test statistic: $z = \dfrac{0.575 - 0.50}{\sqrt{0.5(1 - 0.5)/400}} = 3.00$

4) P-value: 0.003

5) Conclusion: We can reject the null hypothesis at a significance level of 0.05; we have strong evidence that the population proportion of voters who chose Box is different from 0.50.

b) If the sample size had been 40, the test statistic would have been $z = \dfrac{0.575 - 0.50}{\sqrt{0.5(1 - 0.5)/40}} = 0.95$, and the P-value would have been 0.34. We could not have rejected the null hypothesis under these circumstances.

c) The result of a significance test can depend on the sample size. As the sample size increases, the standard error decreases (because the sample size is the denominator of the standard error equation; dividing by a larger number leads to a smaller result). A smaller standard error leads to a larger z-score and a smaller P-value.

9.71 Tax to reduce global warming?:

1) Assumptions: The data are categorical (yes vs. no); the sample is a random sample of 1002 adults; the expected number of yes and no responses are both at least 15 under the null hypothesis: $np = n(1-p) = (0.5)(1002) \geq 15$.

2) Hypotheses: H_0: $p = 0.50$; H_a : $p \neq 0.50$

3) Test statistic: $z = \dfrac{0.32 - 0.50}{\sqrt{0.5(1 - 0.5)/1002}} = -11.4$

4) P-value: 0.000

5) Conclusion: We can reject the null hypothesis using a significance level of 0.05 since the P-value is much smaller than 0.05. We have very strong evidence that less than half of the population think the government should increase taxes on gasoline so people either drive less or buy cars that use less gas.

9.73 Ellsberg paradox:

a) p = proportion of people who would pick Box A; Hypotheses: H_0: $p = 0.50$; H_a : $p \neq 0.50$

b) Test statistic: $z = \dfrac{0.9 - 0.5}{\sqrt{0.5(1 - 0.5)/40}} = 5.06$; P-value$\approx 0$; there is very strong evidence that the population proportion that chooses Box A is not 0.50. Rather, it appears to be much higher than 0.50. Box A seems to be preferred.

9.75 Interest charges on credit card:

The output first tells us that we are testing "p=0.50 vs not =0.50." That tells us that the null hypothesis is that the two cards are preferred equally and the alternative hypothesis is that one is preferred more than the other. The printout then tells us that X is 40. This is the number of people in the sample who preferred the card with the annual cost. 100 is the "N," the size of the whole sample. The "Sample p" of 0.40000 (rounds to 0.40) is the proportion of the sample that preferred the card with the annual cost. The "95.0% CI" is the 95% confidence interval, the range of plausible values for the population proportion. The "Z-Value" is the test statistic. The sample proportion is 2.00 standard errors below the proportion as per the null hypothesis, 0.50. Finally, the "P-Value" tells us that if the null hypothesis were true, the proportion 0.0455 of samples would fall at least this far from the null hypothesis proportion of 0.50. This is barely extreme enough to reject the null hypothesis with a significance level of 0.05. We have evidence that the population proportion of people who prefer the card with the annual cost is below 0.50. The majority of the customers seem to prefer the card without the annual cost, but with the higher interest rate.

9.77   **Type I and Type II errors**:
a)   In the previous exercise, a Type I error would have occurred if we had rejected the null hypothesis, concluding that women were being passed over for jury duty, when they really were not. A Type II error would occur if we had failed to reject the null, but women really were being picked disproportionate to their representation in the jury pool.
b)   If we made an error, it was a Type I error.

9.79   **Practice steps of test for mean**:
a)   Software verifies all of these numbers.
b)   The P-value of 0.002 is less than the significance level of 0.05. We can reject the null hypothesis. We have very strong evidence that the population mean is not 0.
c)   If we had used the one-tailed test, $H_a : \mu > 0$, the P-value would be 0.001, also less than the significance level of 0.05. Again, we have very strong evidence that the population mean is positive.
d)   If we had used the one-tailed test, $H_a : \mu < 0$, the P-value would be 0.999, far from the significance level of 0.05. It would be plausible that the null hypothesis is correct; we cannot conclude that the population mean is negative.

9.81   **Hours at work**:
a)   Hypotheses: $H_0 : \mu = 40; H_a : \mu \neq 40$
b)   (i)   SE Mean, 0.272, is the standard error.
    (ii)   T = 7.64 is the test statistic.
    (iii)   The P-value, 0.000, is the probability of obtaining a sample mean at least this far from the value in H_0 if the null hypothesis were true. In this case, this probability is close to zero.
c)   The test using a significance level of 0.05 indicates that we have strong evidence that people work more than 40 hours per week. The confidence interval supports this conclusion because 40 is below the range of plausible values.

9.83   **Blood pressure**:
a)   1)   Assumptions are that the data are quantitative, have been produced randomly, and have an approximate normal population distribution.
    2)   Hypotheses: $H_0 : \mu = 130; H_a : \mu \neq 130$
    3)   The sample mean is 150.0, and the standard deviation is 8.37
        Test statistic: $t = \dfrac{(150 - 130)}{8.37 / \sqrt{6}} = 5.85$
    4)   P-value: 0.002
    5)   Conclusion: If the null hypothesis were true, the probability would be 0.002 of getting a test statistic at least as extreme as the value observed. There is very strong evidence that the population mean is different from 130; we can conclude that Vincenzo Baronello's blood pressure is not in control.
b)   The assumptions are outlined in Step 1 in part (a). Blood pressure readings are quantitative data. These data are the last six times he monitored his blood pressure. This might be considered a random sample of possible readings for that point in time. We do not know whether the population distribution is normal, but the two-sided test is robust for violations of this assumption.

9.85   **Tennis balls in control?**:
a)   Software indicates a test statistic of -5.5 and a P-value of 0.001.
b)   For a significance level of 0.05, we would conclude that the process is not in control. The machine is producing tennis balls that weigh less than they are supposed to.
c)   If we rejected the null hypothesis when it is in fact true, we have made a Type I error and concluded that the process is not in control when it actually is.

9.87 **Wage claim false?**:
1) Assumptions: The data are quantitative. The data seem to have been produced using randomization. We also assume an approximately normal population distribution.
2) Hypotheses: $H_0: \mu = 500$; $H_a: \mu < 500$
3) The sample mean is 441.11, and the standard deviation is 12.69.

   Test statistic: $t = \dfrac{(441.11 - 500)}{12.69/\sqrt{9}} = -13.9$
4) P-value: 0.000
5) Conclusion: If the null hypothesis were true, the probability would be almost 0 of getting a test statistic at least as extreme as the value observed. There is extremely strong evidence that the population mean is less than 500; we can conclude that the mean income is less than $500 per week.

9.89 **CI and test connection**:
a) We can reject the null hypothesis.
b) It would be a Type I error.
c) A 95% confidence interval would not contain 100. When a value is rejected by a test at the 0.05 significance level, it does not fall in the 95% confidence interval.

9.91 **How to reduce chance of error?**:
a) The researcher can control the probability of a Type I error by choosing a smaller significance level. This will decrease the probability of a Type I error.
b) If a researcher sets the probability equal to 0.00001, the probability of a Type I error is low, but it will be extremely difficult to reject the null hypothesis, even if the null hypothesis is not true.

9.93 **P(Type II error) with smaller n**:
a) $\sqrt{0.333(1-0.333)/60} = 0.061$
b) When P(Type I error) = significance level= 0.05, $z = 1.645$, and the value 1.645 standard errors above 0.333 is 0.433.
c) $\sqrt{0.5(1-0.5)/60} = 0.0645$

   $z = (0.433 - 0.5)/0.065 = -1.03$

   The probability that $\hat{p}$ falls below this z-score is 0.15. The Type II error is larger when n is smaller, because a smaller n results in a larger standard error and makes it more difficult to have a sample proportion fall in the rejection region. If we're less likely to reject the null with a given set of proportions, we're more likely to fail to reject the null when we should reject it.

CHAPTER PROBLEMS: CONCEPTS AND INVESTIGATIONS

🖥 **9.95** **Class data**:
The report will be different for each student.

9.97 **Baseball home team advantage**:
a) These data give us a sense of what the probability would look like in the long run. If we look at just a few games, we don't get to see the overall pattern, but when we look at a number of games over time, we start to see the long run probability of the home team winning.

b) $z = \dfrac{0.5463 - 0.5}{\sqrt{0.5(1-0.5)/2429}} = 4.6$.

   P-value = 0.000
   We can add and subtract the result of (1.96)(0.0101), namely the z-score associated with a 95% confidence interval multiplied by the standard error, to the sample proportion to obtain a confidence interval of (0.53, 0.57) for the probability of the home team winning. The test merely indicates whether $p = 0.50$ is plausible whereas the confidence interval displays the range of plausible values.

9.99   **Two-sided or one-sided?**:

a) Once a researcher sees the data, he or she knows in which direction the results lie. At this point, it is "cheating" to decide to a do a one-tailed test. In this scenario, one has actually done a two-tailed test, then cut the P-value in half upon seeing the results, making it easier to reject the null hypothesis. The decision of what type of test to use must be made before seeing the data

b) A result that is statistically significant with a P-value of 0.049 is not greatly different from one that is not statistically significant with a P-value of 0.051. The decision to use such a cutoff is arbitrary, is dependent on sample size, and is dependent on the random nature of the sample. Such a policy leads to the inflation of significant findings in the public view. If there really is no effect, but many studies are conducted, eventually someone will achieve significance, and then the journal will publish a Type I error.

9.101   **Subgroup lack of significance**:

The sample size (n) has an impact on the P-value. The subgroups have smaller sample size, so for a particular size of effect will have a smaller test statistic and a larger P-value.

9.103   **Overestimated effect**:

The studies with the most extreme results will give the smallest P-values and be most likely to be statistically significant. If we could look at how results from all studies vary around a true effect, the most extreme results would be out in a tail, suggesting an effect much larger than it actually is.

9.105   **Why not accept H_0?**:

When we do not reject H_0, we should not say that we accept H_0. Just because the sample statistic was not extreme enough to conclude that the value in H_0 is unlikely doesn't mean that the value in H_0 is the actual value. As a confidence interval would demonstrate, there is a whole range of plausible values for the population parameter, not just the null value.

9.107   **Significance**:

Statistical significance means that we have strong evidence that the true parameter value is either above or below the value in H_0; this need not indicate practical significance. Practical significance means that the true parameter is sufficiently different from the value in H_0 to be important in practical terms. Examples will vary.

9.109   **Medical diagnosis error**:

With the probability of a false positive diagnosis being about 50% over the course of 10 mammograms, it would not be unusual for a woman to receive a false positive over the course of having had many mammograms. Likewise, when conducting many significance tests with a type I error of 0.05, it would not be unusual to have some show statistical significance (i.e., support the alternative hypothesis) even though the null hypothesis is in fact true.

9.111   **Interpret P-value:**

The P-value tells us the probability of getting a test statistic this large if the value at the null hypothesis represents the true parameter. In this case, it is 0.057. We can reject the null hypothesis if the P-value is at or beyond the significance level α. If it were any number below this, there would not be enough evidence.

9.113   **Small P-value:**

The best answer is (b).

9.115   **Pollution:**

The best answer is (a).

9.117   **True or false:**

False.

9.119 **True or false 3**:
False.

9.121 **True or false 5**:
False.

♦♦**9.123 Standard error formulas**:
If the sample probability is 0, then the standard error is 0, and the test statistic is infinity, which does not make sense. A significance test is conducted by supposing the null is true, so in finding the test statistic we should substitute the null hypothesis value, giving a more appropriate standard error.

CHAPTER PROBLEMS: STUDENT ACTIVITIES

9.125 The results will be different for each class.

SECTION 10.1: PRACTICING THE BASICS

10.1 **Wealth gap:**
   a) The response variable is net worth and the explanatory variable is race.
   b) The two groups that are the categories of the explanatory variable are white and black households.
   c) The samples of white and black households were independent. No household could be in both samples.

10.3 **Binge drinking:**
   a) The estimated difference between the population proportions in 2005 and 1993 is 0.07. The proportion of students who reported bingeing at least 3 times within the past 2 weeks has apparently increased between 1993 and 2005.
   b) The standard error is the standard deviation of the sampling distribution of differences between the sample proportions.

$$se = \sqrt{\frac{\hat{p}_1(1-\hat{p}_1)}{n_1} + \frac{\hat{p}_2(1-\hat{p}_2)}{n_2}} = \sqrt{\frac{(0.312)(0.688)}{159} + \frac{0.382(0.618)}{485}} = 0.0429$$

   c) $0.07 - (1.96)(0.0429) = -0.01$
   $0.07 + (1.96)(0.0429) = 0.15$
   $(-0.01, 0.15)$
   We can be 95% confident that the population mean change in proportion is between -0.01 and 0.15. This confidence interval contains zero; thus, we do not have enough evidence to conclude that there was an increase in the population proportion of UW students who reported binge drinking at least 3 times in the past 2 weeks between 1993 and 2005.
   d) The assumptions are that the data are categorical (reported binge drinking at least 3 times in the past 2 weeks vs. did not), that the samples are independent and are obtained randomly, and that there are sufficiently large sample sizes. Specifically, each sample should have at least ten "successes" and ten "failures."

💻10.5 **Do you believe in miracles?:**
   a) $\hat{p}_1 = 0.454$, and $\hat{p}_2 = 0.563$

   b) $se = \sqrt{\frac{\hat{p}_1(1-\hat{p}_1)}{n_1} + \frac{\hat{p}_2(1-\hat{p}_2)}{n_2}} = \sqrt{\frac{(0.454)(1-0.454)}{500} + \frac{(0.563)(1-0.563)}{622}} = 0.030$

   $0.109 - (1.96)(0.030) = 0.05$
   $0.109 + (1.96)(0.030) = 0.17$
   $(0.05, 0.17)$
   We can be 95% confident that the population proportion for females falls between 0.05 and 0.17 higher than the population proportion for males. Because 0 does not fall in this interval, we can conclude that females are more likely than are males to say that they believe in religious miracles. The assumptions are that the data are categorical, that the samples are independent and obtained randomly, and that there are sufficiently large sample sizes.
   c) The confidence interval has a wide range of plausible values for the population mean difference in proportions, including some (such as 0.05) that indicate a small difference and some (such as 0.17) that indicate a relatively large one.

10.7   **Swedish study test:**

a) $H_0: p_1 = p_2 ; H_a : p_1 \neq p_2$

b)   The P-value of 0.14 tells us that, if the null hypothesis were true, we would obtain a difference between sample proportions at least this extreme 0.14 of the time.

c)   The P-value would be 0.07.

d)   The bigger the sample size, the smaller the standard error and the bigger the test statistic. This study has smaller samples than the Physicians Health Study did. Therefore, its standard error was larger and its test statistic was smaller. A smaller test statistic has a larger P-value.

10.9   **Drinking and unplanned sex:**

a)   Assumptions: Each sample must have at least ten outcomes of each type. The data must be categorical, and the samples must be independent random samples.
Notation: p is the probability that someone says that he/she had engaged in unplanned sexual activities because of drinking alcohol.

$H_0: p_1 = p_2 ; H_a : p_1 \neq p_2$

b) $\hat{p} = (38+146)/(159+485) = 184/644 = 0.286$; this is the common value of p_1 and p_2, estimated by the proportion of the *total* sample who reported that they had engaged in such activities.

c) $se_0 = \sqrt{\hat{p}(1-\hat{p})\left(\dfrac{1}{n_1}+\dfrac{1}{n_2}\right)} = \sqrt{0.286(1-0.286)\left(\dfrac{1}{159}+\dfrac{1}{485}\right)} = 0.04.$

In this case, the standard error is interpreted as the standard deviation of the estimates ($\hat{p}_1 - \hat{p}_2$) from different randomized studies using these sample sizes.

d) $z = \dfrac{(\hat{p}_1 - \hat{p}_2)-0}{se_0} = \dfrac{0.24-0.30}{0.04} = -1.5.$

P-value= 0.13; if the null hypothesis were true, the probability would be 0.13 of getting a test statistic at least as extreme as the value observed. We have insufficient evidence to reject the null hypothesis; we are unable to show that there is a difference in proportions of reports of engaging in unplanned sexual activities because of drinking between 1993 and 2005.

10.11   **Hormone therapy for menopause:**

a)   Assumptions: Each sample must have at least ten outcomes of each type. The data must be categorical, and the samples must be independent random samples.
Notation: p is the probability that someone developed cancer.

$H_0: p_1 = p_2 ; H_a : p_1 \neq p_2$

b)   The test statistic is 1.03, and the P-value is 0.303 (rounds to 0.30). If the null hypothesis were true, the probability would be 0.30 of getting a test statistic at least as extreme as the value observed. It is plausible that the null hypothesis is correct and that there are the results for the hormone therapy group are not different from the results for the placebo group.

c)   We cannot reject the null hypothesis.

10.13   **Living poorly:**
The samples are independent. There is no reason to believe that the responses for those sampled in Japan are related to the responses for those sampled in the U.S.

SECTION 10.2: PRACTICING THE BASICS

10.15 **Address global warming:**
   a) The response variable is the amount of tax the student is willing to add to gasoline in order to encourage drivers to drive less or to drive more fuel-efficient cars; the explanatory variable is whether the student believes that global warming is a serious issue that requires immediate action or not.
   b) Independent samples; the students were randomly sampled so which group the student falls in (yes or no to second question) should be independent of the other students.
   c) A 95% confidence interval for the difference in the population mean responses on gasoline taxes for the two groups, $\mu_1 - \mu_2$, is given by $(\bar{x}_1 - \bar{x}_2) \pm t_{.025}(se)$ where $\bar{x}_1$ is the sample mean response on the gasoline tax for the group who responded 'yes' to the second question, $\bar{x}_2$ is the sample mean response on the gasoline tax for the group who responded 'no' to the second question, t is the t-score for a 95% confidence interval and $se = \sqrt{\dfrac{s_1^2}{n_1} + \dfrac{s_2^2}{n_2}}$ is the standard error of the difference in mean responses.

10.17 **More confident about housework:**
   a) The margin of error is the t value of 2.576 multiplied by the standard error of $0.297 = 0.8$. The bounds are $14.5 - 0.8 = 13.7$, and $14.5 + 0.8 = 15.3$.
   b) This interval is wider than the 95% confidence interval because we have chosen a larger confidence level, and thus, the t value associated with it will be higher. To be more confident, we must include a wider range of plausible values.

▣10.19 Ideal number of children:
   a) The 413 males who responded had a mean of 2.92 and a standard deviation of 1.71.
   b) The 95% confidence interval for the difference between the population means for females and males is

$$se = \sqrt{\frac{s_1^2}{n_1} + \frac{s_2^2}{n_2}} = \sqrt{\frac{(1.92)^2}{470} + \frac{(1.71)^2}{413}} = 0.12216.$$

The margin of error is 1.96 multiplied by the standard error of 0.12216 is 0.24. The confidence interval is the mean difference plus and minus the margin of error: $0.28 - 0.24 = 0.04$, and $0.28 + 0.24 = 0.52$. The confidence interval is (0.04, 0.52). Because 0 is not in this interval, there may be a difference in the population mean reports of ideal number of children between males and females.

🖳10.21 Test comparing auction prices:
   a)

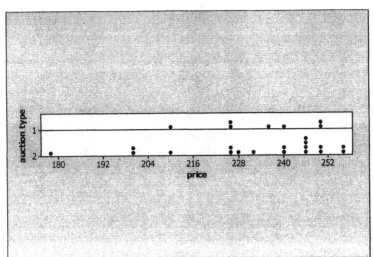

   The dot plot tells us that the bidding only method leads to more variable prices than the "buy it now" auctions. Also, there's an apparent outlier in the bidding only method (i.e., 178).
   b)   The test statistic is 0.26, and the P-value is 0.80. If the null hypothesis were true, the probability would be 0.80 of getting a test statistic at least as extreme as the value observed. It is plausible that the null hypothesis is correct and that there is no difference in population mean prices between auction methods.
   c)   The assumptions are that the data are quantitative, both samples are independent and random, and there is approximately a normal population distribution for each group. The distribution for the bidding only prices may be skewed to the left. Because of the robustness of the two-sided test with respect to this assumption, however, this does not likely affect the validity of our analysis greatly.

10.23   Some smoked but didn't inhale:
   a)   (i)   The overwhelming majority of non-inhalers must have had HONC scores of 0 because the mean is very close to 0 (and there's a small standard deviation). It would only be this low with a large number of scores of 0.
        (ii) On the average, those who reported inhaling had a mean score that was 2.9 – 0.1 = 2.8 (rounds to 3) higher than did those who did not report inhaling.
   b)   The HONC scores were probably not approximately normal for the non-inhalers. The lowest possible value of 0, which was very common, was only a fraction of a standard deviation below the mean.
   c) $se = \sqrt{\dfrac{s_1^2}{n_1} + \dfrac{s_2^2}{n_2}} = \sqrt{\dfrac{(3.6)^2}{237} + \dfrac{(0.5)^2}{95}} = 0.24$

   The standard error is interpreted as the standard deviation of the difference between sample means from different studies using these sample sizes.
   d)   Because 0 is not in this interval, we can conclude that there is a difference in population mean HONC scores between inhalers and non-inhalers at the 95% confidence level. Inhalers appear to have a higher mean HONC score than non-inhalers do.

10.25   Females or males more nicotine dependent?:
   a) $se = \sqrt{\dfrac{s_1^2}{n_1} + \dfrac{s_2^2}{n_2}} = \sqrt{\dfrac{(3.6)^2}{150} + \dfrac{(2.9)^2}{182}} = 0.36$

   The standard error is the standard deviation of the difference between sample from different studies using these sample sizes.

109

b) $t = \dfrac{(\bar{x}_1 - \bar{x}_2) - 0}{se} = \dfrac{(2.8 - 1.6) - 0}{0.364} = 3.30$

P-value: 0.001

If the null hypothesis were true, the probability would be 0.001 of getting a test statistic at least as extreme as the value observed. We have very strong evidence that there is a difference between men's and women's population mean HONC scores. The females seem to have a higher population mean HONC score than the males.

c) The HONC scores were probably not normal for either gender. The standard deviations are bigger than the means, an indication of skew. The lowest possible value of 0 is 2.8/3.6 = 0.778 standard deviations below the mean for females, and 1.6/2.9 = 0.552 standard deviations below the mean for males, indicating skew to the right in both cases. This does not affect the validity of our inference greatly because of the robustness of the two-sided test for the assumption of a normal population distribution for each group.

⌨10.27 TV watching and gender:

a) $H_0: \mu_1 = \mu_2 ; H_a: \mu_1 \neq \mu_2$

b) The test statistic is 1.26 and the P-value is 0.21. If the null hypothesis were true, the probability would be 0.21 of getting a test statistic at least as extreme as the value observed. With a significance level of 0.05, we cannot reject the null hypothesis. Because the probability is 0.21 of observing our test statistic or one more extreme, due to random variation, we have insufficient evidence that there is a gender difference in TV watching.

c) Yes, since we failed to reject the null hypothesis of no difference between the population means, 0 is a plausible value for this difference.

d) The distribution of TV watching is not likely normal. The standard deviations are almost as large as the means (the lowest possible value of 0 is 2.99/2.34 = 1.28 standard deviations below the mean for females and 2.86/2.22 = 1.29 standard deviations below the mean for males), an indication of skew to the right. This does not affect the validity of our inference greatly because of the robustness of the two-sided test for the assumption of a normal population distribution for each group. Inferences assume a randomized study and normal population distributions.

⌨10.29 Study time:

a) Let group 1 represent the students who planned to go to graduate school and group 2 represent those who did not. Then, $\bar{x}_1 = 11.67$, $s_1 = 8.34$, $\bar{x}_2 = 9.10$ and $s_2 = 3.70$. The sample mean study time per week was higher for the students who planned to go to graduate school, but the times were also much more variable for this group.

b) $se = \sqrt{\dfrac{8.34^2}{21} + \dfrac{3.70^2}{10}} = 2.16$. If further random samples of these sizes were obtained from these populations, the differences between the sample means would vary. The standard deviation of these values would equal about 2.2.

c) A 95% confidence interval is (-1.9, 7.0). We are 95% confident that the difference in the mean study time per week between the two groups is between -1.9 and 7.0 hours. Since 0 is contained within this interval, we cannot conclude that the population mean study times differ for the two groups.

10.31 Time spent on e-mail:

a) Let group 1 represent males and group 2 represent females. Then, $\bar{x}_1 = 2.91$, $s_1 = 2.91$, $\bar{x}_2 = 3.10$ and $s_2 = 5.52$. The sample mean time spent on e-mail was slightly higher for females than for males, but notice the apparent outlier for the female group (25). The data were also much more variable for females, but this may also merely reflect the outlier.

b) $se = \sqrt{\dfrac{2.91^2}{11} + \dfrac{5.52^2}{20}} = 1.51$. If further random samples of these sizes were obtained from these populations, the differences between the sample means would vary. The standard deviation of these values would equal about 1.5.

c) A 90% confidence interval is (-2.8, 2.4). We are 90% confident that the difference in the population mean number of hours spent on e-mail per week is between -2.8 and 2.4 for males and females Since 0 is contained within this interval, we cannot conclude that the population mean time spent on e-mail per week differs for males and females.

10.33 Normal assumption:
With large random samples, the sampling distribution of the difference between two sample means is approximately normal regardless of the shape of the population distributions. Substituting sample standard deviations for unknown population standard deviations then yields an approximate t sampling distribution. With small samples, the sampling distribution is not necessarily bell-shaped if the population distributions are highly non-normal.

SECTION 10.3: PRACTICING THE BASICS

10.35 Bulimia test:
a) $t = ((2.0 - 4.8) - 0)/1.025 = -2.73$
P-value: 0.011
b) The assumptions are that the data are quantitative, constitute random samples from two groups, and are from populations with approximately normal distributions. In addition, we assume that the population standard deviations are equal. Given the large standard deviations of the groups, the normality assumption is likely violated, but we're using a two-sided test, so inferences are robust to that assumption.

10.37 Comparing clinical therapies:
a) From software, choosing "assume equal variances,"
C1 N Mean StDev SE Mean
1 3 40.00 8.66 5.0
2 3 20.0 10.0 5.8
95% CI for difference: (-1.2054, 41.2054)
Both use Pooled StDev = 9.3541
b) We can be 95% confident that the population mean difference between change scores is between -1.2 and 41.2. Because 0 falls in this range, it is plausible that there is no difference in mean change scores between the two populations. We do not have sufficient evidence to conclude that there is a mean difference between the two populations. The therapies may not have different means, but if they do the population mean could be much higher for therapy 1. The confidence interval is so wide because the two sample sizes are very small.
c) When we conduct this analysis again with a 90% confidence interval, we get:
90% CI for difference: (3.7178, 36.2822)
Both use Pooled StDev = 9.3541
At this confidence level, we can conclude that therapy 1 is better. 0 is no longer in the range of plausible values.

10.39 Vegetarians more liberal?:
a) The first set of inferences assumes equal population standard deviations, but the sample standard deviations suggest this is not plausible. It is more reliable to conduct the second set of inferences, which do not make this assumption.
b) Based on the first set of results, we would not conclude that the population means are different. 0 falls within the confidence interval; thus, it is plausible that there is no population mean difference. Moreover, the P-value is not particularly small. Based on the second set of results, however, we would conclude that the population means are different. 0 is not in the confidence interval, and hence, is not a plausible value. In addition, the P-value is quite small, an indication that results such as these would be very unlikely if the null hypothesis were true. From this set of results, it appears that the vegetarian students are more liberal than are the non-vegetarian students.

10.41 Fish and heart disease:
a) The probability of heart disease for women who ate fish 5 or more times per week was estimated to be 0.66 times the probability of heart disease for women who ate fish less than once a month.
b) The confidence interval gives us the range of plausible values for the reported relative risk. Because the interval does not contain 1, in the population we can infer that the probability of heart disease was smaller for women who ate fish 5 or more times per week.

10.43 Obesity and cancer:
In the population of women we are 95% confident that the increase in the population death rate for the highly obese group was between <u>40%</u> and <u>87%</u>.

10.45 We're getting heavier:
The ratio of means is 191/166 = 1.15. The weight of men in 2002 was estimated to be 1.15 the weight of men in 1962. The men increased by 15%.

SECTION 10.4: PRACTICING THE BASICS

10.47 Does exercise help blood pressure?:
a) The three "before" observations and the three "after" observations are dependent samples because the same patients are in both samples.
b) The sample mean of the "before" scores is 150, of the "after" scores is 130, and of the difference scores is 20. The difference between the means for the "before" and "after" scores is the same as the mean of the difference scores.
c) From software, the standard deviation of the difference scores is 5.00.

$se = s_d / \sqrt{n} = 5/\sqrt{3} = 2.887$; For a 95% confidence interval given by $\bar{x}_d \pm t_{.025}$ (se), the lower endpoint is $20 - (4.303)(2.887) = 7.6$ and the upper endpoint is $20 + (4.303)(2.887) = 32.4$. Thus, the 95% confidence interval is (7.6, 32.4).

We can be 95% confident that the difference between the population means is between 7.6 and 32.4. Because 0 is not included in this interval and because all differences are positive, we can conclude that there is a decrease in blood pressure after patients go through the exercise program.

10.49 Social activities for students:
a) To compare the mean movie attendance and mean sports attendance using statistical inference, we should treat the samples as dependent; the same students are in both samples.
b) There is quite a bit of spread, but the outliers appear in both directions.
c) The 95% confidence interval was obtained by adding and subtracting the margin of error from the sample mean difference score of 4.0. The margin of error is calculated by multiplying the appropriate t-score of 2.262 (for $df=9$) by the standard error of the difference scores of 5.11. The margin of error, therefore, is 11.6.
d) The test statistic was obtained as: $t = \dfrac{\bar{x}_d - 0}{se} = \dfrac{4.0 - 0}{5.11} = 0.78$

The P-value is 0.45. It is plausible that the null hypothesis is true. Because the P-value is large, we cannot conclude that there is a population mean difference in attendance at movies vs. sports events.

🖳10.51 Movies vs. parties:
a) 1) The assumptions made by these methods are that the difference scores are a random sample from a population distribution that is approximately normal.
2) $H_0: \mu_1 = \mu_2$ (or the population mean of difference scores is 0); $H_a: \mu_1 \neq \mu_2$
3) $t = -1.62$
4) P-value: 0.140 (rounds to 0.14)
5) If the null hypothesis were true, the probability would be 0.14 of getting a test statistic at least as extreme as the value observed. Because the probability is 0.14 of observing a test statistic or one more extreme by random variation, we have insufficient evidence that there is a mean population difference between attendance at movies vs. attendance at parties.

    b) (-21.111, 3.511) which rounds to (-21.1, 3.5); we are 95% confident that the population mean number of times spent attending movies is between 21.1 less and 3.5 higher than the population mean number of times spent attending parties.

10.53   **Checking for "freshmen 15":**
    a) The standard deviation of the change in weight scores could be much smaller because even though there is a lot of variability among the initial and final weights of the women, most women do not see a large change in weight over the course of the study, so the weight changes would not vary much.
    b) $se = s_d / \sqrt{n} = 2.0 / \sqrt{110} = 0.191$; A 95% confidence interval given by $\overline{x}_d \pm t_{.025}$ (se) has lower endpoint $0.8 - (1.982)(0.191) = 0.4$ and upper endpoint $0.8 + (1.982)(0.191) = 1.2$. Thus, the confidence interval is (0.4, 1.2). 15 is not a plausible weight change in the population of Freshmen women. The plausible weight change falls in the range from 0.4 to 1.2.
    c) The data must be quantitative, the sample of difference scores must be a random sample from a population of such difference scores, and the difference scores must have a population distribution that is approximately normal (particularly with samples of size less than 30).

⌨10.55 Comparing book prices 2:
    1) Assumptions: the differences in prices are a random sample from a population that is approximately normal.
    2) H_0: $\mu_d = 0$ vs. H_a: $\mu_d \neq 0$.
    3) $t = \dfrac{\overline{d} - 0}{s_d / \sqrt{n}} = \dfrac{4.3}{4.7152 / \sqrt{10}} = 2.88$.
    4) P-value=0.02.
    5) If the null hypothesis is true, the probability of obtaining a difference in sample means as extreme as that observed is 0.02. We would reject the null hypothesis and conclude that there is a significant difference in prices of textbooks used at her college between the two sites for $\alpha=0.05$ or 0.10, but not for $\alpha=0.01$.

10.57   **Treat juveniles as adults?:**
    a) They are dependent since matched pairs were formed by matching certain criteria and one juvenile from each pair was assigned to either juvenile court or adult court.
    b) Let group 1 represent the juveniles assigned to adult court and group 2 represent the juveniles assigned to juvenile court. Then $\hat{p}_1 = 673/2097 = 0.32$ and $\hat{p}_2 = 448/2097 = 0.21$.
    c) 1) Assumptions: the difference in re-arrest rates are a random sample from a population that is approximately normal.
       2) H_0: $\mu_d = 0$ vs. H_a: $\mu_d \neq 0$.
       3) $z = \dfrac{515 - 290}{\sqrt{515 + 290}} = 7.9$.
       4) P-value≈0.
       5) If the null hypothesis is true, the probability of obtaining a difference in sample means as extreme as that observed is very close to 0. There is extremely strong evidence of a population difference in the re-arrest rates between juveniles assigned to adult court and those assigned to juvenile court.

10.59   **Change coffee brand?:**
    a) The point estimates for the population proportions choosing Sanka for the first and second purchases, respectively, are $204/541 = 0.38$ and $231/541 = 0.43$.
    b) Each proportion in (a) can be found as a sample mean by assigning Sanka a "1" and Other brand a "0". For each subject, their coffee score is 1 if they choose Sanka, and 0 if they choose another brand. The mean of the scores is the proportion using Sanka. The estimated difference of population proportions is the difference between the sample means at the two times: $0.38 - 0.43 = -0.05$.
    c) The confidence interval of (0.01, 0.09) tells us that we can be 95% confident that the population proportion choosing Sanka for the second purchase was between 0.01 and 0.09 higher than the population proportion choosing it for the first purchase.

10.61 **Heaven and hell**:
  a) The point estimate for the difference between the population proportions believing in heaven and believing in hell is $0.631 - 0.496 = 0.135$.
  b) (i) The assumptions are that the data are categorical, the samples are independent and random, and the sum of the two counts in the test is at least 30.

   (ii) $H_0: p_1 = p_2$; $H_a: p_1 \neq p_2$

   (iii) $z = \dfrac{65 - 35}{\sqrt{65 + 35}} = 3.00$

   (iv) P-value: 0.003

   (v) This is a very small P-value; if the null hypothesis were true, the probability would be 0.003 of getting a test statistic at least as extreme as the value observed. We have strong evidence that there is a difference between the population proportions of those who believe in heaven and those who believe in hell. It appears that more people believe in heaven than in hell.

SECTION 10.5: PRACTICING THE BASICS

10.63 **Benefits of drinking**:
  a) This refers to an analysis of three variables, a response variable, an explanatory variable, and a control variable. The response variable is "whether have heart disease;" the explanatory variable is "whether drink alcohol moderately;" and the control variable is age.
  b) There is a stronger association between drinking alcohol and whether have heart disease for subjects who are older (in fact, perhaps no association for young subjects).

10.65 **Basketball paradox**:
  a) The proportion of shots made is higher for Barry both for 2-point shots and for 3-point shots, but the proportion of shots made is higher for O'Neal overall.
  b) O'Neal took almost exclusively 2-point shots, where the chance of success is higher.

10.67 **Family size in Canada**:
  a) The mean number of children for English-speaking families (1.95) is higher than the mean number of children in French-speaking families (1.85).
  b) Controlling for province, this association reverses. In each case, there's a higher mean for French-speaking families. For Quebec, the mean number for French-speaking families (1.80) is higher than for English-speaking families (1.64). Similarly, for other provinces, the mean for French-speaking families (2.14) is higher than the mean for English-speaking families (1.97).
  c) This paradox likely results from the fact that there are relatively more English-speaking families in the "other" provinces that tend to produce more children regardless of language, and more French-speaking families in Quebec where they tend to have fewer children regardless of language. This illustrates Simpson's paradox.

10.69 **Breast cancer over time**:
There could be no difference in the prevalence of breast cancer now and in 1900 for women of a given age. Overall, the breast cancer rate would be higher now, because more women live to an old age now, and older people are more likely to have breast cancer.

CHAPTER PROBLEMS: PRACTICING THE BASICS

10.71 **Opinion about America**:
  a) The response variable is view of the U.S., and the explanatory variable is year.
  b) Separate samples of subjects should be treated as independent in order to conduct inference. The two samples are not linked (e.g., not the same people in both groups), and hence, are independent of each other.
  c) Now, the two samples include the same subjects, and so should be treated as dependent.

10.73   Marijuana and Gender:
a) We are 95% confident that the population proportion of females who have used marijuana is at least 0.0077 lower and at most 0.0887 lower than the population proportion of males who have used marijuana. Because 0 is not in the confidence interval, we can conclude that females and males differ with respect to marijuana use.
b) The confidence interval would change only in sign. It would now be (0.0077, 0.0887). We are 95% confident that the population proportion of males who have used marijuana is at least 0.0077 higher and at most 0.0887 higher than the population proportion of females who have used marijuana.

10.75   Belief depend on gender?:

a) $se_0 = \sqrt{\hat{p}(1-\hat{p})\left(\dfrac{1}{n_1}+\dfrac{1}{n_2}\right)} = \sqrt{0.8048(1-0.8048)\left(\dfrac{1}{625}+\dfrac{1}{502}\right)} = 0.0238$

b) $z = \dfrac{(\hat{p}_1 - \hat{p}_2) - 0}{se_0} = \dfrac{0.0216 - 0}{0.0238} = 0.91;$ P-value: 0.3628 (which rounds to 0.36).

If the null hypothesis were true, the probability would be 0.36 of getting a test statistic at least as extreme as the value observed. This P-value is not smaller than a typical significance level such as 0.05; therefore, we cannot reject the null hypothesis. It is plausible that the null hypothesis is correct and that the population proportions believing in an afterlife are the same for females and males.

c) If the population difference were $0.81 - 0.74 = 0.07$, our decision would have been in error. There actually would have been a difference between females and males.

d) The assumptions on which the methods in this exercise are based are that we used independent random samples for the two groups and that we had at least 10 successes and 10 failures in each sample.

10.77   Heavier horseshoe crabs more likely to mate?:
a)

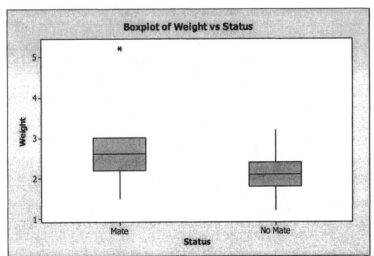

The female crabs have a higher median and a bigger spread if they had a mate than if they did not have a mate. The distribution for the female crabs with a mate is right-skewed, whereas the distribution for the female crabs without a mate is symmetrical.

b) The estimated difference between the mean weights of female crabs who have mates and who do not have mates is $2.6 - 2.1 = 0.5$.

c) $se = \sqrt{\dfrac{s_1^2}{n_1}+\dfrac{s_2^2}{n_2}} = \sqrt{\dfrac{0.36}{111}+\dfrac{0.16}{62}} = 0.076$

d) $(\overline{x}_1 - \overline{x}_2) \pm t_{.05}(se)$

Because n_1 and n_2 are large, we will approximate t with the normal distribution using $z = 1.645$.
$0.5 - (1.645)(0.076) = 0.375$
$0.5 + (1.645)(0.076) = 0.625$
$(0.375, 0.625)$
We can be 90% confident that the difference between the population mean weights of female crabs with and without a mate is between 0.375 and 0.625 kg. Because 0 does not fall in this interval, we can conclude that female crabs with a mate weigh more than do female crabs without a mate.

10.79 Test TV watching by race:
a) $H_0: \mu_1 = \mu_2; H_a: \mu_1 \neq \mu_2$
b) The test statistic is 8.08, and the P-value is 0.000. If the null hypothesis were true, the probability would be almost 0 of getting a test statistic at least as extreme as the value observed.
c) Using a significance level of 0.05, we can reject the null hypothesis. We have strong evidence that there is a difference in mean TV watching time between the populations of blacks and whites. Blacks seem to watch more TV than do whites.
d) When we reject the null hypothesis at a significance level of 0.05, a 95% confidence interval does not include the value at the null hypothesis – in this case, 0.

10.81 Time spent on Internet:
a) The response variable is the number of hours a week spent on the Internet and is quantitative. The explanatory variable is the respondent's gender and is categorical.
b) The 99% confidence interval is (0.4, 2.2). We are 99% confident that the population mean number of hours spent on the Internet per week is between 0.4 and 2.2 hours more for males than for females.
c) 1) Assumptions: Independent random samples, and number of hours spent on the Internet per week has an approximately normal population distribution for each gender.
   2) $H_0: \mu_1 = \mu_2; H_a: \mu_1 \neq \mu_2$.
   3) $t = 3.62$.
   4) P-value=0.000.
   5) If the null hypothesis is true, the probability of obtaining a difference in sample means as extreme as that observed is close to 0. At $\alpha=0.01$, we would reject the null hypothesis and conclude that the population mean number of hours a week spent on the Internet differs for males and females.

10.83 Sex roles:
1) Assumptions: the data are quantitative (child's score); the samples are independent and we will assume that they were collected randomly; we assume that the population distributions of scores are approximately normal for each group.
2) $H_0: \mu_1 = \mu_2; H_a: \mu_1 \neq \mu_2$ where group 1 represents the group with the male tester and group 2 represents the group with the female tester.
3) $se = \sqrt{\dfrac{1.4^2}{50} + \dfrac{1.2^2}{90}} = 0.2349$, $t = \dfrac{2.9 - 3.2}{0.2349} = -1.28$.
4) P-value: 0.205
5) If the null hypothesis were true, the probability would be 0.205 of getting a test statistic at least as extreme as the value observed. Since the P-value is quite large, there is not much evidence of a difference in the population mean of the children's scores when the tester is male versus female.

10.85 Parental support and household type:
a) The 95% confidence interval is 4 plus and minus 3.4, that is (0.6, 7.4).
b) The conclusion refers to the results of a significance test in that it tells us the P-value of 0.02. If the null hypothesis were true, the probability would be 0.02 of getting a test statistic at least as extreme as the value observed.

10.87   Teenage anorexia:

a) The P-value of 0.10 indicates that if the null hypothesis that there is no difference between population mean change scores were true, the probability would be 0.10 of getting a test statistic at least as extreme as the value observed.

b) The assumptions for these analyses are that the data are quantitative, the samples are independent and random, and the population distributions for each group are approximately normal. Based on the box plots (which show outliers and a skew to the right for the cognitive-behavioral group), it would not be a good idea to conduct a one-sided test. It is not as robust as the two-sided test to violations of the normal assumption.

c) The lowest plausible difference between population means is the lowest endpoint of the confidence interval, -0.7, a difference of less than 1 pound. Thus, if there is a change in this direction (cognitive-behavioral group less), then it is less than 1 pound. On the other hand, the highest plausible difference between population means is the highest endpoint of the confidence interval, 7.6. Thus, if there is a change in this direction (cognitive-behavioral group more), then it could be almost as much as 8 pounds.

d) The confidence interval and test give us the same information. We do not reject the null hypothesis that the difference between the population means is 0, and 0 falls in the 95% confidence interval for the difference between the population means.

⌨10.89 Surgery vs. placebo for knee pain:

Sample N Mean StDev SE Mean
1   60 48.9 21.9   2.8
2   59 51.7 22.4   2.9
Difference = mu (1) - mu (2)
Estimate for difference: -2.80000
95% CI for difference: (-10.84257, 5.24257)
T-Test of difference = 0 (vs not =): T-Value = -0.69 P-Value = 0.492 DF = 117
Both use Pooled StDev = 22.1493
The confidence interval is (-10.8, 5.2). We can be 95% confident that the population mean difference falls in this interval. Because 0 falls in this range, it is plausible that there is no difference between the population mean pain scores of these two groups.

10.91   Anorexia again:

a) We can be 95% confident that the population mean difference is between -0.7 and 7.6. Because 0 falls in this range, it is plausible that there is no population mean difference between these groups.

b) The P-value of 0.10 indicates that if the null hypothesis were true, the probability would be 0.10 of getting a test statistic at least as extreme as the value observed.

c) P-value = 0.10/2 = 0.05. If the null hypothesis were true, the probability would be 0.05 of getting a test statistic of 1.68 or larger. This P-value is smaller than the one in (b), thus providing stronger evidence against the null hypothesis.

d) For these inferences, we assume a quantitative response variable, independent random samples, and approximately normal population distributions for each group.

10.93   Eat tomatoes:

a) The relative risk of 0.77 indicates that, in this study, for those who ate tomato sauce, the sample proportion of prostate cancer was estimated to be 0.77 times the sample proportion of prostate cancer cases for those who did not eat tomato sauce.

b) We can be 95% confident that the population mean relative risk value is between 0.66 and 0.90. Because 1.0 does not fall in this range, all of the plausible values indicate a decreased risk among those who eat tomato sauce.

10.95   Prison rates:

a) The ratio of incarceration rates is (1/109)/(1/1563)=14.3. Males were 14.3 times more likely to be incarcerated than women in 2006.

b) The ratio of incarceration rates is (1694/100000)/(252/100000)=6.7. Black residents were 6.7 times more likely to be incarcerated than white residents in 2006.

10.97 Improving employee evaluations:
We would explain that there's less than a 5% chance that we'd get a sample mean at least this much higher after the training course if there were, in fact, no difference in the population. To make this more informative, it would have been helpful to have the sample means or the sample mean difference and its standard error – or better yet, the confidence interval.

10.99 Effect of alcoholic parents:
a) The groups are dependent since they were matched according to age and gender.
b) 1) Assumptions: the differences in scores are a random sample from a population that is approximately normal.
2) H_0: $\mu_d = 0$ vs. H_a: $\mu_d \neq 0$.
3) $t = \dfrac{2.7}{9.7/\sqrt{49}} = 1.95$.
4) P-value=0.057.
5) If the null hypothesis is true, the probability of obtaining a difference in sample means at least as extreme as that observed is 0.057. This is some, but not strong, evidence that there is a difference in the mean scores between children of alcoholics versus children of non-alcoholics.
c) We assume that the population of differences is approximately normal and that our sample is a random sample from this distribution.

10.101 Breast augmentation and self esteem:
a) The samples were dependent since the same women were sampled before and after their surgeries.
b) No. In order to find the *t* statistic, we need to know the standard deviation of the differences which cannot be obtained from the information given.

⌨10.103 TV or rock music a worse influence?:
a) The samples are dependent; the same people are answering both questions.
b) According to software:
95% CI for mean difference: (-0.335, 1.335) which rounds to (-0.3, 1.3); we can be 95% confident that the population mean difference between ratings of the influence of TV and rock music is between -0.3 and 1.3. Because 0 falls in this range, it is plausible that there is no difference between the population mean ratings.
c) According to software:
T-Test of mean difference = 0 (vs not = 0): T-Value = 1.32 P-Value = 0.21; the P-value of 0.21 indicates that if the null hypothesis were true, the probability would be 0.21 of getting a test statistic at least as extreme as the value observed.

10.105 Crossover study:
a) high dose: 69/86 = 0.802; low dose: 61/86 = 0.709
b) $z = \dfrac{16 - 8}{\sqrt{16 + 8}} = 1.63$; P-value: 0.10; if the null hypothesis were true, the probability would be 0.10 of getting a test statistic at least as extreme as the value observed. It is plausible that the null hypothesis is correct and that there is no difference between low-dose and high-dose analgesics with respect to population proportions who report relief of menstrual bleeding.
The sum of the counts in the denominator should be at least 30, although in practice the two-sided test works well even if this is not true. In addition, the sample should be an independent random sample, and the data should be categorical.

10.107 Fast food, TV, and obesity:
The "triple risk" refers to the ratio of proportions for the two groups; that is, the proportion of obese children group in the group that eats fast food and watches lots of TV is three times the proportion of obese children in the group that does not.

10.109 Death penalty paradox:

a) When we ignore victim's race, we observe a proportion of 0.11 white defendants who receive the death penalty, and a proportion of 0.079 (rounds to 0.08) black defendants who receive the death penalty. It appears that whites are more likely to receive the death penalty than are blacks.

When we take victim's race into account, the direction of the association changes with black defendants more likely to get the death penalty.

Specifically, when the victim was white: White defendants have a probability of 0.113 (rounds to 0.11) and blacks have a probability of 0.229 (rounds to 0.23) of receiving the death penalty.

When the victim was black: White defendants have a probability of 0.000 (rounds to 0.00) and blacks have a probability of 0.028 (rounds to 0.03) of receiving the death penalty.

In both cases, the proportion of blacks who receive the death penalty is higher than the proportion of whites who receive the death penalty.

b) The death penalty was imposed more frequently when the victim was white, and white victims where more common when the defendant was white.

10.111 Income and gender:

a) The mean income difference could disappear if we controlled for number of years since receiving highest degree. If most of the female faculty had been hired recently, they would be fewer years from their degree, and would have lower incomes. So, the overall mean could be lower for females. If we look only at those who are a given year from receiving their degree (e.g., received degree five years ago), we might find no gender difference.

b) The mean income difference could disappear if we controlled for college of employment. If more women seek positions in low salary colleges and more men in high salary colleges, it might appear that men make more. If we look only within a given college (e.g., law school), we might not find a gender difference in income.

CHAPTER PROBLEMS: CONCEPTS AND INVESTIGATIONS

10.113 More eBay selling prices:

The one-page summaries will be different for each student, but should include findings such as those in the following software output.

Sample N   Mean StDev SE Mean
1   136 223.4 23.7   2.0
2   132 235.5 20.6   1.8
Difference = mu (1) - mu (2)
Estimate for difference: -12.0900
95% CI for difference: (-17.4281, -6.7519)
T-Test of difference = 0 (vs not =): T-Value = -4.46 P-Value = 0.000 DF = 262

10.115 Pay discrimination against women?:

a) We would need to know the sample standard deviations and sample sizes for the two groups.

b) It would not be relevant to conduct a significance test. A significance test lets us make inferences about a population based on a sample. If we already have the information on the entire population, there's no need to make inferences about the population.

10.117 Obesity and earnings:

a) Whether obese (yes or no) and wage are stated to have an association.

b) Education level is one possibility. The women could be paired according to education level and then compared in obesity rates.

10.119 Comparing mean incomes:

The best answer is (d).

10.121 Positive values in CI:
False.

10.123 Control for clinic:
False.

♦♦10.125 Standard error of difference:

$$se(\text{estimate 1} - \text{estimate 2}) = \sqrt{[se(\text{estimate 1})]^2 + [se(\text{estimate 2})]^2} = \sqrt{(0.6)^2 + (1.8)^2} = 1.897$$

$(\bar{x}_1 - \bar{x}_2) \pm 1.96(se)$

$(46.3 - 33.0) - 1.96(1.897) = 9.58$ (round to 9.6)
$(46.3 - 33.0) + 1.96(1.897) = 17.02$ (round to 17.0)
$(9.6, 17.0)$

We can be 95% confident that the difference between the population mean number of years lost is between 9.6 and 17.0. Because 0 is not in this range, we can conclude that there is a population mean difference between those who smoke and are overweight and those who do not smoke and are normal weight in terms of number of years of life left. It appears that those who do not smoke and are normal weight have more years left than do those who smoke and are overweight.

♦♦10.127 Small-sample CI:

a) (i) $\hat{p}_1 = \hat{p}_2 = 0$ because there are no successes in either group (i.e., $0/10 = 0$).

    (ii) $se = 0$ because there is no variability in either group if all responses are the same. Specifically, both numerators under the square root sign in the se formula would have a zero in them, leading to a calculation of 0 as the se.

    (iii) The 95% confidence interval would be (0,0) because we'd be adding 0 to 0 (se multiplied by z would always be 0 with se of 0, regardless of the confidence level).

b) Using the small-sample method:

    (i) For each group $\hat{p} = 1/12 = 0.083$

    (ii) $se = \sqrt{\dfrac{0.083(1-0.083)}{12} + \dfrac{0.083(1-0.083)}{12}} = 0.113$

$(\hat{p}_1 - \hat{p}_2) \pm z(se)$

$(0.083 - 0.083) - (1.96)(0.113) = -0.221$ (rounds to -0.22)
$(0.083 - 0.083) + (1.96)(0.113) = 0.221$ (rounds to 0.22)
The new confidence interval, (-0.22, 0.22), is far more plausible than (0, 0).

CHAPTER PROBLEMS: STUDENT ACTIVITIES

10.129 Reading the medical literature:
The reports will differ based on the article chosen by the class instructor.

REVIEW PROBLEMS: PRACTICING THE BASICS

R3.1 **Reincarnation:**

  a) $se = \sqrt{\hat{p}(1-\hat{p})/n} = \sqrt{0.27(1-0.27)/2201} = 0.0095$.

  b) The standard error would be twice as large ($\sqrt{4} = 2$). In order to increase the precision of estimates, the standard error must get smaller which happens when the sample size increases. Since the sample size is in the denominator through its square root, it must quadruple for the standard error to be half as large.

R3.3 **Homosexual relations:**

  a) The point estimate is given by 0.54. A 95% confidence interval is given by $\hat{p} \pm 1.96se$ where

  $se = \sqrt{\hat{p}(1-\hat{p})/n}$. Thus, a 95% confidence interval is $0.54 \pm 1.96\sqrt{0.54(1-0.54)/1200}$
  $=0.54\pm0.028$, or (0.51, 0.57).

  b) 1) Assumptions: The variable, whether sexual relations between two adults of the same sex is wrong, is categorical. The samples in the two years are independent random samples. The sample sizes, 1200 each, are large enough to insure that the sampling distribution of the difference between the sample proportions is approximately normal. [Check that the number of successes and failures for each group are greater than 5.]

  2) Hypotheses: $H_0: p_1 = p_2 ; H_a: p_1 \neq p_2$

  3) Test Statistic: $z = 0.20/0.0196 = 10.2$.

  4) P-value ≈ 0.

  5) Conclusion: Since the P-value is smaller than the significance level of 0.05, we reject the null hypothesis and conclude that the proportion of Floridians who say sexual relations between two adults of the same sex is always wrong has changed from years 1988 to 2006. There is extremely strong evidence that it has decreased

R3.5 **Reduce Services, or Raise Taxes?:**

  a) 1) Assumptions: The variable, whether to raise taxes or reduce services, is categorical. The sample is a random sample. The sample size, 1200, is large enough to insure that the sampling distribution of the sample proportion is approximately normal. [Check that the number of successes and failures are both greater than15.]

  2) Let p be the proportion of adult Floridians who favor raising taxes to handle the problem. Hypotheses: $H_0: p=0.5$ versus $H_a: p \neq 0.5$.

  3) Test Statistic: $z = \dfrac{\hat{p} - p_0}{\sqrt{p_0(1-p_0)/n}} = \dfrac{0.52-0.5}{\sqrt{0.5(1-0.5)/1200}} = 1.39$.

  4) P-value$=2P(z > 1.39)=0.16$.

  5) Conclusion: Since the P-value is not very small, there is not much evidence about whether a majority or minority of Floridians favored raising taxes to handle the government's problem of not having enough money to pay for all of its services.

  b) $n = \dfrac{\hat{p}(1-\hat{p})z^2}{m^2} = \dfrac{0.52(1-0.52)1.96^2}{0.03^2} = 1065$.

R3.7 **Same-sex marriage in Canada:**

a) The hypotheses are H_0: $p=0.5$ versus H_a: $p \neq 0.5$. The assumptions are as follows: (i) the variable, whether the bill legalizing same-sex marriage should stand, is categorical; (ii) the sample is a random sample; (iii) the sample size, 1000, is large enough to insure that the sampling distribution of the sample proportion is approximately normal. [Check that the number of successes and failures are both greater than15.]

b) Test Statistic: $z = \dfrac{\hat{p} - p_0}{\sqrt{p_0(1 - p_0)/n}} = \dfrac{0.55 - 0.5}{\sqrt{0.5(1 - 0.5)/1000}} = 3.16$. The sample proportion, 0.55, lies 3.16 standard errors above the hypothesized value of 0.5.

c) P-value=$2P(z > 3.16)=0.002$. If the null is true, the probability of obtaining a sample result at least as extreme as that observed is about 0.002.

d) Since the P-value is very small, there is very strong evidence that in 2005 a majority of Canadians thought that the bill legalizing same-sex marriage should stand.

R3.9 **Sex partners:**

a) $se = s/\sqrt{n} = 17.1/\sqrt{227} = 1.13$.

b) We are 95% confident that in 2004 the population mean number of male sex partners for females between the ages of 20 and 29 was between 4.9 and 9.3.

c) Since the standard deviation is more than twice the mean, the distribution of number of male sex partners for females between the ages of 20 and 29 is probably highly skewed to the right. Note that the smallest value, 0, is only $7.1/17.1 = 0.4$ standard deviations below the mean. Since the sample size is quite large, the confidence interval is still valid because the sampling distribution will still be bell-shaped by the Central Limit Theorem.

d) The median may be a more appropriate measure of center since the population distribution is likely to be very highly skewed.

⌨R3.11 Legal marijuana?:

a) The samples are independent since the respondents differ from one year to the next.

b) The percentage favoring legalization showed a significant increase in the 70s but dipped back down in the 80s. The percentage in favor of legalization has been increasing fairly steadily since 1991.

R3.13 **Laughter and blood flow:**

The samples are dependent since the same 20 people were observed watching both of the films.

R3.15 **Listening to rap music:**

a) We are 95% confident that the population proportion of black youth who listen to rap music every day is between 0.09 and 0.17 higher than the population proportion of Hispanic youths who listen to rap music every day.

b) In order for it to be plausible that the population proportions are identical for black and Hispanic youths, 0 would need to be included in the interval.

R3.17 **Evolution:**

a) Let p_1 denote the proportion of fundamentalists who answered 'definitely not true' and p_2 denote the proportion of liberals who answered 'definitely not true'. Then the hypotheses are $H_0 : p_1 = p_2$ versus $H_a : p_1 \neq p_2$.

b) Test statistic: $z = \dfrac{(\hat{p}_1 - \hat{p}_2) - 0}{se_0}$ with $se_0 = \sqrt{\hat{p}(1-\hat{p})\left(\dfrac{1}{n_1} + \dfrac{1}{n_2}\right)}$ where $\hat{p} = \dfrac{190+60}{323+309} = 0.3956$. Thus,

$$z = \dfrac{\left(190/323 - 60/309\right)}{\sqrt{0.3956(1-0.3956)\left(\dfrac{1}{323} + \dfrac{1}{309}\right)}} = 10.13.$$

c) P-value=2P(z > 10.13) ≈ 0.

d) Since the P-value is approximately 0, it is highly unlikely that we would obtain a test statistic as extreme as that observed if the null hypothesis were true. We conclude that the population proportion who responded 'definitely not true' when asked if human beings evolved from earlier species of animals is higher for those who classify themselves as religious fundamentalists than for those who classify themselves as liberal in their religious beliefs.

R3.19   No time cooking:

A 95% confidence interval is given by $(\hat{p}_1 - \hat{p}_2) \pm 1.96(se)$ where $se = \sqrt{\dfrac{\hat{p}_1(1-\hat{p}_1)}{n_1} + \dfrac{\hat{p}_2(1-\hat{p}_2)}{n_2}}$. For this

example, $se = \sqrt{\dfrac{0.45(0.55)}{1219} + \dfrac{0.26(0.74)}{733}} = 0.0216$. A 95% confidence interval for the difference between the

population proportions of men and women who reported spending no time on cooking and washing up during a typical day is then given by (0.45-0.26)±1.96(0.0216), or (0.15, 0.23). We are 95% confident that the population proportion of men who reported spending no time on cooking and washing up during a typical day is between 0.15 and 0.23 higher than the population proportion of women who responded the same.

R3.21   Compulsive buying:
a) 1) Assumptions: the response, whether the respondent is a compulsive buyer, is categorical for both groups; the groups are independent and the samples are random; the sample sizes are large enough so that there are at least five successes and five failures for each group.
    2) Let group 1 be the sample of women and group 2 the sample of men. Then the hypotheses are
        $H_0 : p_1 = p_2$ versus $H_a : p_1 \neq p_2$.
    3) Test statistic: $z = 0.48$.
    4) P-value =0.63.
    5) The P-value is quite large indicating that the test statistic observed is not unusual under the null hypothesis. We are unable to conclude that there is a difference in the population proportions of male and female compulsive buyers.
b) A 95% confidence interval is given by (-0.015, 0.025). Since 0 is contained in the interval, we are unable to conclude that there is a difference in the population proportions of males and females who are compulsive buyers.

R3.23   Housework for women and men:
a) The 95% confidence interval is given by (11.9, 16.1). We are 95% confident that the population mean number of minutes per day spent on cooking and washing up is between 11.9 and 16.1 minutes more for women than for men. Since 0 does not fall within this interval, we can conclude that the mean time spent per day on cooking and washing up is higher for women than for men.
b) The assumptions are that the samples are random and independent and that the number of minutes per day spent on cooking and washing up is approximately normally distributed for the two groups.

R3.25 Loneliness:
a) For males, group 1, $\bar{x}_1 = 1.28$, for females, group 2, $\bar{x}_2 = 1.67$.
b) A 95% confidence interval is given by $\bar{x}_2 \pm 1.96se = 1.67 \pm 1.96\left(2.30/\sqrt{816}\right) = (1.51, 1.83)$. We can be 95% confident that the mean number of days in the past 7 days that females have felt lonely is between 1.5 and 1.8.

c) A 95% confidence interval is given by $\left(\bar{x}_1 - \bar{x}_2\right) \pm t_{.025} se$. The confidence interval for the difference

between males and females is $(1.28-1.67)\pm1.96\sqrt{\dfrac{2.06^2}{634}+\dfrac{2.30^2}{816}}$ =(-0.6, -0.2). We can be 95% confident

that the population mean number of days that males felt lonely over the past 7 days is between 0.6 and 0.2 days less than women felt lonely. Since 0 is not contained in the confidence interval, we can conclude that for a given week males felt lonely less than women on average (but the difference could be quite small).

R3.27 Sex partners and gender:
a) The P-value, 0.000, is the probability of obtaining a test statistic at least as extreme as that observed if the null hypothesis is true. Since the P-value is close to 0, there is extremely strong evidence that the population mean number of sex partners differs for males and females.
b) We are 95% confident that the population mean number of sex partners over the past year is between 0.53 and 0.25 less for females than males. The confidence interval not only tells us that there is a significant difference (0 is not contained in the interval), but it also gives us an idea of what the population mean difference is.
c) We assume that the samples are independent, random samples from distributions that are approximately normal.

R3.29 Binge eating:
a) The difference in the proportion of women who suffer from binge eating versus suffer from anorexia is 0.035 − 0.01= 0.025.
b) The relative risk is given by (0.035/0.01) 3.5. Thus, the proportion of women who suffer from binge eating is 3.5 times the proportion of women who suffer from anorexia.

R3.31 Improving math scores:
a) (i) $\bar{x}_{after} = \dfrac{70+80+\ldots+97}{10} = 78.0$, $\bar{x}_{before} = \dfrac{60+73+\ldots+96}{10} = 71.0$; thus,

$\bar{x}_{after} - \bar{x}_{before} = 78-71=7$. (ii) To find the mean of the difference scores, we must first calculate the differences: 70-60=10, 80-73=7, 40-42= -2, 94-88=6, 79-66=13, 86-77=9, 93-90=3, 71-63=8, 70-55=15,

97-96=1. Then, $\bar{x}_d = \dfrac{10+7+\ldots+1}{10} = 7.0$. The two methods give identical answers.

b) $t = \dfrac{\bar{x}_d - 0}{s_d / \sqrt{n}} = \dfrac{7}{5.25 / \sqrt{10}} = 4.216$. Thus, P-value=2P($t > 4.216$)=0.002. The probability of obtaining a

test statistic at least as extreme as that observed, assuming the null hypothesis is true, is 0.002. Since the P-value is quite small, there is very strong evidence of a difference in the population mean scores before and after the training course.

c) A 90% confidence interval is given by $\bar{x}_d \pm t_{.05}\left(s_d / \sqrt{n}\right) = 7 \pm 1.83\left(5.25 / \sqrt{10}\right) = (4.0, 10.0)$. We are

90% confident that the population mean difference in scores after and before taking the training course is between 4.0 and 10.0. Since 0 is not included in the interval, we conclude that the population mean test score was higher after the training course.

d) The 90% confidence interval contains 0. Likewise, the significance test has P-value below 0.10. Each inference suggests that 0 is not a plausible value for the population mean difference.

REVIEW PROBLEMS: CONCEPTS AND INVESTIGATIONS

⊟R3.33 Student survey:

Answers will vary but could include a significance test and/or confidence interval. The test steps are:

1) Assumptions: the response, weekly number of times reading the newspaper, is quantitative; the samples are random and independent; the weekly number of times reading a newspaper is approximately normally distributed for each of the two groups (although this assumption is not met, the sample sizes are relatively large and we are conducting a two-sided test which is robust to violations of this assumption).

2) Let group 1 be females and group 2 be males. Then, the hypotheses are $H_0 : \mu_1 = \mu_2$ versus

$H_a : \mu_1 \neq \mu_2$.

3) Test statistic: $t = \dfrac{(\bar{x}_1 - \bar{x}_2) - 0}{se} = -0.82$

4) P-value=0.42.

5) Since the P-value is quite large, the test statistic observed is not unusual under the assumption of the null hypothesis. We are unable to conclude that the weekly number of times a person reads the newspaper depends on the person's gender.

A 95% confidence interval for the mean difference in weekly number of times reading a newspaper for females versus males is (-2.2, 0.9).

R3.35   More people becoming isolated?:

The proportion of people who said that they had not discussed matters of importance with anyone over the last six months was 0.089 in 1985 and 0.246 in 2004. To make an inference concerning whether these proportions are statistically different, we can construct a confidence interval. A 95% confidence interval is given by

$$\left(\hat{p}_1 - \hat{p}_2\right) \pm z \sqrt{\frac{\hat{p}_1\left(1 - \hat{p}_1\right)}{n_1} + \frac{\hat{p}_2\left(1 - \hat{p}_2\right)}{n_2}} =$$

$$\left(0.089 - 0.246\right) \pm 1.96 \sqrt{\frac{0.089(1 - 0.089)}{1531} + \frac{0.246(1 - 0.246)}{1482}} = \left(-0.18, -0.13\right).$$ Since 0 is not

contained in the confidence interval, we can conclude that the population proportion of people who said that they had not discussed matters of importance with anyone over the last 6 months was less in 1985 than in 2004.

R3.37   Variability and inference:

By the sample size formula $\dfrac{\sigma^2 z^2}{m^2}$, the needed sample size is proportional to the squared standard deviation.

To estimate the mean income for all lawyers in the U.S., we would need a large sample because of the wide range of salaries (large standard deviation) due to differences in specialty, experience, location, etc. To estimate the mean income for all entry-level employees at Burger King restaurants in the U.S., pay is likely to be fairly homogeneous (small standard deviation) so that a small sample would suffice.

R3.39   Freshman weight gain:

If we assume no difference in mean weight gain for the population of freshman men and women, the probability of obtaining a sample difference as large as or larger than that observed would be quite small. This probability is called the P-value and when it is small enough, it contradicts the statement of no difference, providing us with sufficient evidence to reject this statement and conclude the two groups have differing mean weight gains.

R3.41   Overweight teenagers:

Yes, an increase of 12% in the percentage of teenagers who are overweight seems practically significant as well. Statistical significance means that the sample results were 'extreme' enough to reject the null hypothesis of no difference. The results are practically significant if they have practical value, this must be determined by the investigator.

♦♦R3.43 Comparing literacy:

Although the sample sizes are needed to calculate the *t*-score used in the confidence interval, a *z*-score can be used when the sample sizes are sufficiently large and similiar. Assuming that this is the case, a 95% confidence interval comparing the difference in population means for Canada and the U.S. is given by

$$(286.9 - 277.9) \pm 1.96\sqrt{3^2 + 2^2}$$, which is (1.9, 16.1). We are 95% confident that the difference in population mean prose literacy scores for Canada and the U.S. is between 1.9 and 16.1. Since 0 is not contained in the interval, we can conclude that the population mean prose literacy score is higher in Canada than in the U.S.

♦♦R3.45 Margins of error:

a) The margin of error is given by $z\sqrt{\dfrac{\hat{p}(1-\hat{p})}{n}} = 1.96\sqrt{\dfrac{0.6(0.4)}{1000}} = 0.03$. The limits for either 40% or 60% are 3.0 points.

b) By the formula given in part (a), we see that the margin of error will change for different sample proportions, and tends to get smaller as the sample proportion moves toward 0 or 1.

c) Since the numerator for the standard error contains the term $\hat{p}(1-\hat{p})$, the margin of error will be the same for a particular sample proportion and for 1 minus that value.

Chapter 11

Analyzing the Association between Categorical Variables

SECTION 11.1: PRACTICING THE BASICS

11.1 **Gender gap in politics?:**

a) The response variable is political party identification, and the explanatory variable is gender.

b) **POLITICAL PARTY IDENTIFICATION**

GENDER	Democrat	Independent	Republican	Total	*n*
Females	35.8%	40.2%	24.0%	100%	2470
Males	28.3%	44.1%	27.6%	100%	1949

Women are more likely than are men to be Democrats, whereas men are more likely than are women to be Independents or Republicans.

c) There are many possible hypothetical conditional distributions for which these variables would be independent. Distributions should show percentages in the party categories that are the same for men and women. Here's one such example:

POLITICAL PARTY IDENTIFICATION

GENDER	Democrat	Independent	Republican	Total	*n*
Females	37.9%	35.7%	26.4%	100%	2470
Males	37.9%	35.7%	26.4%	100%	1949

d) The second of these two graphs is one possible example in which these variables are independent.

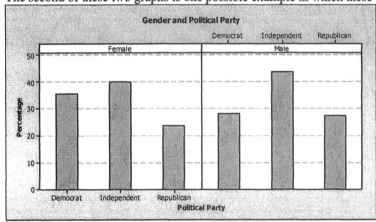

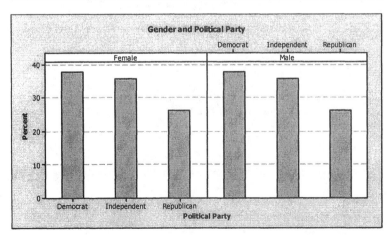

11.3   **FBI statistics:**

a)   These distributions refer to those of x at given categories of y.

	RACE OF VICTIM	
RACE OF MURDERER	**Blacks**	**Whites**
Blacks	91%	17%
Whites	9%	83%

b) x and y are dependent because the probability of a murderer being of a certain race changes greatly according to the race of the person murdered.

⌨**11.5**   **Marital happiness and income:**

a)

	HAPPINESS OF MARRIAGE			
INCOME	**Very Happy**	**Pretty Happy**	**Not Too Happy**	***n***
1-2	155	136	13	304
3	442	256	18	716
4-5	256	115	6	377

b)

	HAPPINESS OF MARRIAGE			
INCOME	**Very Happy**	**Pretty Happy**	**Not Too Happy**	***n***
1-2	51.0	44.7	4.3	304
3	61.7	35.8	2.5	716
4-5	67.9	30.5	1.6	377

Low income people tend to have slightly less happy marriages than those in the other two income brackets, whereas high income people tend to have slightly happier marriages than those in the other two income brackets.

c)   Regardless of income brackets, marital happiness tends to be higher than general happiness.

⌨**11.7**   **Sample evidence about independence:**

In 2006, the percentage in each of the 9 regions was about 31% for very happy, 56% for pretty happy, and 13% for not too happy.   Independence seems plausible. (In fact the chi-squared statistic from the next section equals 9.5, with df = 16, and has a P-value of 0.89.)

Section 11.2: Practicing the Basics

11.9   **Happiness and gender:**

a) H_0: Gender and happiness are independent.

H_a: Gender and happiness are dependent.

b)   The P-value is not especially small. If the null hypothesis were true, the probability would be 0.67 of getting a test statistic at least as extreme as the value observed. So, there is not strong evidence against the null hypothesis, and it is plausible that gender and happiness are independent.

11.11   **Life after death and gender:**

a)

	BELIEF IN LIFE AFTER DEATH		
GENDER	**Yes**	**No**	**Total**
Males	891	233	1124
Females	1286	219	1505

b)

	BELIEF IN LIFE AFTER DEATH		
GENDER	**Yes**	**No**	**Total**
Males	930.75	193.25	1124
Females	1246.25	258.75	1505

There are more women who believe in the afterlife and men who do not than is expected.

c) $X^2 = \sum \dfrac{(\text{observed count - expected count})^2}{\text{expected count}} = (891\text{-}930.75)^2/930.75 + (233\text{-}193.25)^2/193.25 + (1286\text{-}1246.2)^2/1246.2 + (219\text{-}258.75)^2/258.75 = 1.70 + 8.18 + 1.27 + 6.11 = 17.3.$

11.13 **Cigarettes and marijuana**:

a)

	MARIJUANA	
CIGARETTES	**Yes**	**No**
Yes	61.1	38.9
No	5.9	94.1

This conditional distribution suggests that marijuana use is much more common for those who have smoked cigarettes than for those who have not.

b) 1) The assumptions are that there are two categorical variables (cigarette use and marijuana use in this case), that randomization was used to obtain the data, and that the expected count was at least five in all cells.

2) H_0: Cigarette use and marijuana use are independent.

H_a: Cigarette use and marijuana use are dependent.

3) $X^2 = 642.0$

4) P-value: 0.000

5) If the null hypothesis were true, the probability would be close to 0 of getting a test statistic at least as extreme as the value observed. This P-value is quite low. We have extremely strong evidence that marijuana use and cigarette use are associated.

11.15 **Help the environment**:

a) H_0: Sex and desire to help the environment are independent.

H_a: Sex and desire to help the environment are dependent.

b) $r = 2$ and $c = 5$; thus, df $= (r - 1)(c - 1) = (2-1)(5-1) = 4$

c) The P-value is 0.09. It is (i) not less than 0.05, but (ii) is less than 0.10.

d) (i) With a significance level of 0.05, we would fail to reject the null hypothesis; we would conclude that it is plausible that the null hypothesis is correct..

(ii) With a significance level of 0.10, we would reject the null hypothesis; we would conclude that we had strong evidence that desire to help the environment depends on sex.

11.17 **Aspirin and heart attacks**:

a)

	HEART ATTACK		
TREATMENT	**Yes**	**No**	**Total**
Placebo	28	656	684
Aspirin	18	658	676

b) 1) The assumptions are that there are two categorical variables (treatment and heart attack incidence), that randomization was used to obtain the data, and that the expected count was at least five in all cells.

2) H_0: Treatment and incidence of heart attack are independent.

H_a: Treatment and incidence of heart attack are dependent.

3) $X^2 = 2.1$

4) P-value: 0.14

5) If the null hypothesis were true, the probability would be 0.14 of getting a test statistic at least as extreme as the value observed. This is not strong evidence against the null hypothesis. It is plausible that the null hypothesis is correct and that treatment and heart attack incidence are independent.

11.19 Happiness and life after death:

a)

	HAPPINESS			
POSTLIFE	**Very Happy**	**Pretty Happy**	**Not Too Happy**	**Total**
Yes	32.6	54.4	13.0	2175
No	25.8	60.4	13.8	449

People who believe in life after death tend to respond that they are very happy more than those who do not believe in the afterlife whereas those who do not believe in the afterlife are more likely to respond that they are pretty happy than those who do believe in the afterlife. The proportion responding that they are not too happy was similar for both groups.

b) GSS shows a chi square of 7.98 and a P-value of 0.02. If the null hypothesis were true, the probability would be 0.02 of getting a test statistic at least as extreme as the value observed.

c) There is strong evidence that happiness depends on opinion about the afterlife.

11.21 Testing a genetic theory:

a) $H_0 : p = 0.75$ (the probability of a green seedling is 0.75) versus $H_1 : p \neq 0.75$ (the probability of a green seedling is not 0.75).

b) Under the null hypothesis, we would expect 1103(0.75)=827.25 green seedlings and 1103(0.25)=275.75 yellow seedlings. The chi-squared goodness-of-fit statistic is then

$$\chi^2 = \frac{(854 - 827.25)^2}{827.25} + \frac{(249 - 275.75)^2}{275.75} = 3.50 \text{ with } df\text{=2-1=1.}$$

c) P-value=0.06. The probability of obtaining a test statistic as extreme as that observed, assuming the null hypothesis is true, is 0.06. There is evidence against the null, but not very strong.

SECTION 11.3: PRACTICING THE BASICS

11.23 Party ID and race:

a) 1) The assumptions are that there are two categorical variables (party identification and race), that randomization was used to obtain the data, and that the expected count was at least five in all cells.

2) H_0: Party identification and race are independent; H_a: Party identification and race are dependent.

3) $X^2 = 362.1$

4) P-value: 0.000

5) If the null hypothesis were true, the probability would be close to 0 of getting a test statistic at least as extreme as the value observed. We have very strong evidence that party identification depends on race.

b) The proportion of black subjects who identify as Democrats is 394/626 = 0.63; the proportion of white subjects is 859/3208 = 0.27. The difference between these proportions is 0.63 – 0.27 = 0.36. This seems to be a strong association in that it is quite far from 0, the difference corresponding to no association.

11.25 Happiness and highest degree:

a) The P-value associated with a chi-squared statistic of 68.3 is very close to 0. Using a significance level of 0.05, we reject the null hypothesis. It is not plausible that the null hypothesis is correct and that happiness and highest degree are independent.

b) The large chi-squared value does not mean that there is a strong association between happiness and highest degree. It only means that there is strong evidence of an association. Given a large sample size, even a weak association could be statistically significant.

c) 86/440=0.195 of those who did not complete high school say that they're "not too happy". Only 56/794=0.070 in the group who completed college or graduate school fall in this happiness category. This is a difference of 0.13. In the sample, happiness is higher among those who completed college than among those who did not complete high school.

d) The relative risk of being not too happy is 0.195/0.07 = 2.8. People who are college graduates are 2.8 times as likely to be happy as people who did not graduate from high school.

11.27 Smoking and alcohol:

a) 64.02% of those who had not smoked cigarettes also used alcohol, whereas 96.92% of those who had smoked cigarettes had used alcohol. 96.92 – 64.02 = 32.9. The proportion who had used alcohol is 0.33 higher for cigarette users than for non-cigarette users.

b) 14.07% of those who had not used alcohol smoked cigarettes, whereas 74.35% of those who had used alcohol also used cigarettes. 74.35 – 14.07 = 60.28. The proportion who had used cigarettes is 0.60 higher for alcohol users than for those who had not used alcohol.

c) The relative risk of using cigarettes is 0.744/0.141 = 5.3. Alcohol users are 5.3 times as likely to have smoked as are non alcohol users. This does seem to indicate an association between alcohol and cigarette use.

11.29 **Prison and gender**:
a) 832/100,000 = 0.00832 of men were incarcerated, whereas 58/100,000 = 0.00058. The relative risk of being incarcerated is 0.00832/0.00058 = 14.3. Men were 14.3 times as likely as women were to be incarcerated.
b) The difference of proportions being incarcerated is 0.00832 – 0.00058 = 0.008. The proportion of men who are incarcerated is 0.008 higher than the proportion of women who are incarcerated.
c) The response in part (a) is more appropriate because it shows there is a substantial gender effect, which the difference does not show when both proportions are close to 0.

11.31 **Death penalty associations**:
a) False. A larger chi-squared value might be due to a larger sample size rather than a stronger association.
b) It seems that race is more strongly associated with the death penalty opinion, because we can infer that the population difference of proportions is larger than for gender.

SECTION 11.4: PRACTICING THE BASICS

11.33 **Standardized residuals for happiness and income**:
a) The standardized residual indicates that the number of standard errors that the observed count falls from the expected count. In this case, the observed count falls 3.14 standard errors above the expected count.
b) The standardized residuals highlighted in green designate conditions in which the observed counts are much higher than the expected counts, relative to what we'd expect due to sampling variability.
c) The standardized residuals highlighted in yellow designate conditions in which the observed counts are much lower than the expected counts, relative to what we'd expect due to sampling variability.

11.35 **Marital happiness and general happiness**:
a) The relatively small standardized residual of -1.3 indicates that the observed count for this cell is only 1.3 standard errors below the expected count. This is not strong evidence that there is a true effect in that cell.
b) There is less than a 1% chance that a standardized residual would exceed 3 in absolute value, if the variables were independent. Based on this criterion, the following cells would lead us to infer that the population has many more cases than would occur if happiness and marital happiness were independent. People who are not happy in their marriage are more likely to be not happy in general than would be expected if the variables were independent. People who are pretty happy in their marriage are more likely to be pretty happy overall than would be expected if the variables were independent. Lastly, people who are very happy in their marriage are more likely to be very happy overall than would be expected if the variables were independent.

11.37 **Gender gap?**:
There are more women who identify as Democrats, and fewer men who identify as Democrats than would be expected if there were no association between political party and sex. It does seem that political party and sex are not independent – that there is an association.

SECTION 11.5: PRACTICING THE BASICS

11.39 **Keeping old dogs mentally sharp**:
a)

	COULD SOLVE TASK		
CARE AND DIET	**Yes**	**No**	**Total**
Standard	2	6	8
Extra	12	0	12

🖳b)1) There are two binary categorical variables, and randomization was used.

2) H_0: Care and ability to solve the task are independent.

H_a: Care and ability to solve the task are dependent.

3) Test statistic: 2
4) P-value: 0.001
5) If the null hypothesis were true, the probability would be 0.001 of getting a test statistic at least as extreme as the value observed. The P-value is quite low (lower than a significance level of 0.05, for example); we can reject the null hypothesis. We have strong evidence that a dog's care and diet are associated with its ability to solve a task.

c) It is improper to conduct the chi-squared test for these data because the expected cell counts are less than 5 for at least some cells.

11.41   Claritin and nervousness:

a) The P-value for the small-sample test is 0.24. It is plausible that the null hypothesis is true and that nervousness and treatment are independent.

b) It is not appropriate to conduct the chi-squared test for these data because two cells have an expected count of less than five.

CHAPTER PROBLEMS: PRACTICING THE BASICS

11.43   Female for President?:

a)

SEX	Yes	VOTE No	Total
Females	94%	6%	100%
Males	94%	6%	100%

b) If results for the entire population are similar, it does seem possible that gender and opinion about having a woman President are independent. The percentages of men and women who would vote for a qualified woman may be the same.

11.45   Down and chi-squared:

1) The assumptions are that there are two categorical variables (Down Syndrome status and blood test result), that randomization was used to obtain the data, and that the expected count was at least five in all cells.

2) H_0: Down Syndrome status and blood test result are independent.

H_a: Down Syndrome status and blood test result are dependent.

3) $X^2 = 114.4$, df = 1
4) P-value: 0.000
5) If the null hypothesis were true, the probability would be almost 0 of getting a test statistic at least as extreme as the value observed. There is very strong evidence of an association between test result and actual status.

11.47   Echinacea no better than placebo?:

a) The response variable is URI (more than 1 vs. just 1), and the explanatory variable is treatment (placebo vs. Echinacea).

b) I would explain that if having more than one URI did not depend on whether one took Echinacea or placebo, then it would be quite unusual to observe the results actually obtained. This provides relatively strong evidence of a lower rate of URI for those taking Echinacea.

11.49   Gender gap?:

1) The assumptions are that there are two categorical variables (party identification and gender), that randomization was used to obtain the data, and that the expected count was at least five in all cells.

2) H_0: Party identification and gender are independent.

   H_a: Party identification and gender are dependent.

3) $X^2 = 28.8$; df=2
4) P-value: 0.000
5) If the null hypothesis were true, the probability would be very close to 0 of getting a test statistic at least as extreme as the value observed. We have very strong evidence that party identification depends on gender.

⌨11.51 Aspirin and heart attacks:

a) (i)   The assumptions are that there are two categorical variables (group and myocardial infarction), that randomization was used to obtain the data, and that the expected count was at least five in all cells.

   (ii) H_0: Group and myocardial infarction are independent.

        H_a: Group and myocardial infarction are dependent.

   (iii) $X^2 = 26.9$; df=2
   (iv) P-value: 0.000
   (v)   The P-value is very small. If the null hypothesis were true, the probability would be close to 0 of getting a test statistic at least as extreme as the value observed. We have very strong evidence that there is an association between myocardial infarction and group.

b) The proportion of those on placebo who had a fatal heart attack is 0.0017. The proportion of those on aspirin who had a fatal heart attack is 0.0005. Thus, the relative risk is 0.00166/0.00046 = 3.6. Those on placebo are 3.6 times as likely as those on aspirin to have a fatal heart attack.

11.53   Seat belt helps?:

a) The proportion of those who were injured given that they did not wear a seat belt is 0.125. The proportion of those who were injured given that they wore a seat belt is 0.064. The difference between proportions is 0.125 − 0.064 = 0.061. The proportion who were injured is 0.06 higher for those who did not wear a seat belt than for those who did wear a seat belt.

b) The relative risk is 0.125/0.064 = 1.95. People were 1.95 times as likely to be injured if they were not wearing a seat belt than if they were wearing a seat belt.

11.55   Happiness and sex:

It seems as though those with no partners or two or more partners are less likely to be very happy than would be expected if these variables were not associated, and those with 1 partner are more likely to be very happy than would be expected if these variables were independent.

11.57   TV and aggression:

a) H_0: Amount of TV watching and aggression are independent.

   H_a: Amount of TV watching and aggression are dependent.

   p_1 is the population proportion of those who are aggressive in the group that watches less than one hour of TV per day, and p_2 is the population proportion of those who are aggressive in the group that watches more than one hour of TV per day. Then the null can be expressed as $p_1 = p_2$ and the alternative can be expressed as p_1 not equal to p_2.

b) The P-value is 0.0001. This is a very small P-value. If the null hypothesis were true, the probability would be close to 0 of getting a test statistic at least as extreme as the value observed. We have very strong evidence that TV watching and aggression are associated.

CHAPTER PROBLEMS: CONCEPTS AND APPLICATIONS

▣11.59 Student data:

Each student's short report will be different, but could include the following findings.
From MINITAB:
Rows: religiosity   Columns: life_after_death

```
              n       u       y      All

0             8       3       4       15
          3.250   4.000   7.750   15.000

1             5      11      13       29
          6.283   7.733  14.983   29.000

2             0       2       5        7
          1.517   1.867   3.617    7.000

3             0       0       9        9
          1.950   2.400   4.650    9.000

All          13      16      31       60
         13.000  16.000  31.000   60.000


Cell Contents:        Count
                      Expected count

Pearson Chi-Square = 21.386, DF = 6, P-Value = 0.002
```

▣11.61 Another predictor of happiness?:

The one-page report will be different depending on the variable that each student finds to be associated with happiness.

11.63   Babies and gray hair:

a)

	HAS YOUNG CHILDREN	
GRAY HAIR	Yes	No
Yes	0	4
No	5	0

b)

	HAS YOUNG CHILDREN	
GRAY HAIR	Yes	No
Yes	0%	100%
No	100%	0%

There does seem to be an association. All women in the sample who have gray hair do not have young children, whereas all women in the sample who do not have gray hair do have young children.

c)   There often are third factors that influence an association. Gray hair is associated with age (older women being more likely to be gray), and age is associated with having or not having young children (older women being less likely to have young children). Just because two things are associated, doesn't mean that one causes the other.

11.65   Gun homicide in U.S. and Britain:
  a)   The proportion in the U.S. is 0.0000624. The proportion in Britain is 0.0000013. The difference of proportions with the U.S. as group 1 is 0.000061, and with Britain as group 1 is -0.000061. The only thing that changes is the sign.
  b)   The relative risk with the U.S. as group 1 is 0.0000624/0.0000013 = 48. The relative risk with Britain as group 1 is 0.0000013/0.0000624 = 0.021. One value is the reciprocal of the other
  c)   When both proportions are so small, the relative risk is more useful for describing the strength of association. The difference between proportions might be very small, even when one is many times larger than the other.

11.67 $X^2 = 0$:
False.

11.69   Statistical but not practical significance:
True.

11.71   Normal and chi-squared with df = 1:
  a)   The chi squared value for a right-tail probability of 0.05 and df=1 is 3.84, which is the z value for a two-tail probability of 0.05 squared: (1.96)(1.96) = 3.84
  b)   The chi-squared value for P-value of 0.01 and df=1 is 6.64. This is the square of the z value for a two-tail P-value of 0.01, which is 2.58.

♦♦11.73 Standardized residuals for 2x2 tables:
The two observed values in a given row (or column) must add up to the same total as the two expected values in that same row (or column). Thus, we know that if one of these two expected values is above its related observed count, then the other expected value in that same row (or column) must be below its related observed count. For example, let's say that we have a row with observed count of 50 in one cell and an observed count of 50 in the other cell. If the expected count for the first cell is 60, then the expected count for the other must be 40. Both pairs must add up to the same number – in this case, 100.

♦♦11.75 Explaining Fisher's exact test:
  a)   The twenty distinct possible samples that could have been selected are as follows:

F1, F2, F3	F1, F2, M1	F1, F2, M2	F1, F2, M3
F1, F3, M1	F1, F3, M2	F1, F3, M3	F1, M1, M2
F1, M1, M3	F1, M2, M3	F2, F3, M1	F2, F3, M2
F2, F3, M3	F2, M1, M2	F2, M1, M3	F2, M2, M3
F3, M1, M2	F3, M1, M3	F3, M2, M3	M1, M2, M3

This contingency table shows that two males were chosen (M1 and M3) and one was not (M2). It also shows that one female was chosen (F2) and two were not (F1 and F3).
  b)   Ten samples have a difference greater than or equal to 1/3. A difference greater than or equal to 1/3 indicates more men than women in the sample. These samples are as follows:

F1, M1, M2	F1, M1, M3	F1, M2, M3	F2, M1, M2
F2, M1, M3	F2, M2, M3	F3, M1, M2	F3, M1, M3
F3, M2, M3	M1, M2, M3		

CHAPTER PROBLEMS: STUDENT ACTIVITIES

11.77   Conduct a research study using the GSS:
Responses to this exercise will depend on the categorical response variable assigned by the instructor and on the explanatory variables chosen by each student.

Chapter 12
Analyzing Association between Quantitative Variables: Regression Analysis

SECTION 12.1: PRACTICING THE BASICS

12.1 **Car mileage and weight**:
  a) The response variable is mileage, and the explanatory variable is weight.
  b) $\hat{y} = 45.6 - 0.0052x$; the y-intercept is 45.6 and the slope is - 0.0052.
  c) For each 1000 pound increase in the vehicle, the predicted mileage will decrease by 5.2 miles per gallon.
  d) The y-intercept is the predicted miles per gallon for a car that weighs 0 pounds. This is far outside the range of the car weights in this database and, therefore, does not have contextual meaning for these data.

12.3 **Children of working females**:
  a) $\hat{y} = 5.2 - 0.044(28) = 4.0$
  b) $\hat{y} = 5.2 - 0.044(91) = 1.2$
  c) $y - \hat{y} = 2.3 - 1.196 = 1.1$
  d) The y-intercept indicates that for nations with no female economic activity, the predicted fertility rate is 5.2. As x increases from 0 to 100, the predicted fertility rate decreases from 5.2 to 0.8.

12.5 **Mu, not y**:
For a given x-value, there will not be merely one y value because not every elementary schoolgirl in your town who is a given height will weigh the same. It makes more sense to include the mean, μ_y, rather than a specific value, y, in the equation.

⌨12.7 **Study time and college GPA**:
  a)

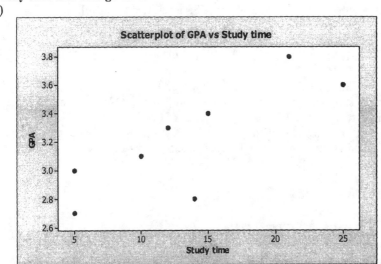

Based on the scatterplot, there appears to be a positive association between GPA and study time.
  b) From MINITAB: `GPA = 2.63 + 0.0439 Study time`. For every 1 hour increase in study time per week, gpa is predicted to increase by about 0.04 points.
  c) predicted GPA=2.63+.0439(25)=3.73.
  d) $y - \hat{y}$ =3.6-3.73= -.13. The observed gpa for student 2, who studies an average of 25 hours per week, is 3.6 which is .13 points below the predicted gpa of 3.73.

⌨**12.9** **Predicting college GPA**:
a) The response variable is college GPA, and the explanatory variable is high school GPA.

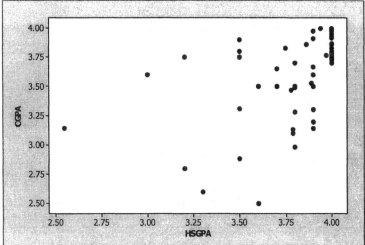

There are several students with high school GPA's of exactly 4.0. This is a ceiling effect that limits our ability to predict college GPA.
b) From MINITAB: predicted `CGPA = 1.19 + 0.637 HSGPA`
(i) CGPA = 1.19 + 0.637 (3.0) = 3.10
(ii) CGPA = 1.19 + 0.637 (4.0) = 3.74; `predicted college GPA increases 0.64 (the slope) for 1-unit increase in high school GPA`.

SECTION 12.2: PRACTICING THE BASICS

12.11 **Dollars and thousands of dollars**:
Slope when income is in dollars: 1.50/1000 = 0.0015

12.13 **When can you compare slopes?**:
a) For a $1000 increase in GDP, the predicted percentage using cell phones increases by 2.62, and the predicted percentage using the Internet increases by 1.55.
b) Because the slope of GDP to cell phone use is larger than is the relation of GDP to Internet use, an increase in GDP would have a slightly greater impact on the percentage using cell phones than on the percentage using the Internet.

12.15 **Sit-ups and the 40-yard dash**:
a) (i) $\hat{y} = 6.7065 - 0.024346(10) = 6.46$

(ii) $\hat{y} = 6.7065 - 0.024346(40) = 5.73$

The difference in predicted times is 6.46 - 5.73 = 0.73 = the slope multiplied by the difference in numbers of sit-ups: (0.024346)(30) = 0.73.
b) Because the slope is positive, the correlation also will be positive.

$$r = b\left(\frac{s_x}{s_y}\right) = 0.024346\left(\frac{6.887}{0.365}\right) = 0.46.$$

12.17 **Student ideology**:
a) This is a negative correlation. That indicates that as people read the newspaper more they tend to become less conservative.
b) $r^2 = (-0.066)(-0.066) = 0.004$. The sum of squared errors is 0.4% less when we use the regression equation instead of using the mean of y. This is a very weak association; both the correlation and the proportional reduction in error are very small.

12.19 **SAT regression toward mean**:
   a) $\hat{y} = 250 + 0.5(800) = 650$
   b) The predicted *y* value will be 0.5 standard deviations above the mean, for every one standard deviation above the mean that *x* is. Here, *x* = 800 is three standard deviations above the mean; so the predicted *y* value is 0.5(3) = 1.5 standard deviations above the mean.

☐12.21 **GPA and study time**:
   a) *r* = .81. There is a fairly strong, positive, linear association between gpa and study time. The more time a student studies, the higher their gpa is likely to be.
   b) $r^2 = 0.656$. The error using $\hat{y}$ to predict y is 66% smaller than the error using $\bar{y}$ to predict y.

12.23 **Placebo helps cholesterol?**:
   a) Their mean cholesterol reading at time 2 should be 200+100(0.7)=270.
   b) This does not suggest that placebo is an effective treatment; this decrease could occur merely because of regression to the mean. Subjects who are relatively high at one time will, on the average, be lower at a later time. So, if a study gives placebo to people with relatively high cholesterol (that is, in the right-hand tail of the blood cholesterol distribution), on the average we expect their values three months later to be lower.

12.25 **What's wrong with your stock fund?**:
   This might be due to regression to the mean. Stocks that are relatively high one year will, on the average, be lower at a later time.

12.27 **Car weight and mileage**:
   There is a 75% reduction in error in predicting a car's mileage based on knowing the weight, compared to predicting by the mean mileage. This relatively large value means that we can predict a car's mileage quite well if we know its weight.

12.29 **Yale and U Conn**:
   The correlation between high-school and college GPA would likely be higher at the University of Connecticut than at Yale. Yale would have a restricted range of high school GPA values, with nearly all of its students clustered very close to the top. U. Conn. would have a wider range of high-school GPAs. The correlation tends to be smaller when we sample only a restricted range of *x*-values than when we use the entire range.

☐12.31 **Correlations for the strong and for the weak**:
   a) From software, the correlation between number of bench presses before fatigue (BRTF(60)) and maximum bench press (1RMBENCH) is 0.80, a strong, positive association.
   b) (i) The median is 10. Using only the *x*-values below 10, the correlation is 0.67.
       (ii) Using only the *x*-values above the median of 10, the correlation is 0.48.
       They are so different because the correlation usually is smaller in absolute value when the range of predictor values is restricted.

SECTION 12.3: PRACTICING THE BASICS

12.33 **Predicting house prices:**
a) i) <u>Assumptions</u>: Assume randomization, linear trend with normal conditional distribution for y and the same standard deviation at different values of x.

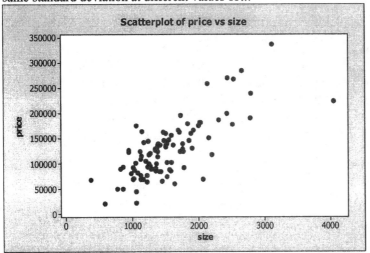

   ii) <u>Hypotheses</u>: The null hypothesis that the variables are independent is H_0: $\beta = 0$. The two-sided alternative hypothesis of dependence is H_a: $\beta \neq 0$.
   iii) <u>Test statistic</u>: According to software, the test statistic is 11.62. We also could calculate the test statistic as follows: $t = b/se = 77.008/6.626 = 11.6$.
   iv) <u>P-value</u>: From software, the P-value is 0.000.
   v) <u>Conclusion</u>: If H_0 were true that the population slope $\beta = 0$, it would be extremely unusual – the probability would be almost 0 – to get a sample slope at least as far from 0 as $b = 77.008$. The P-value gives very strong evidence that an association exists between the size and price of houses; this is extremely unlikely to be due to random variation.

b) The 95% confidence interval is $b \pm t_{.025}(se) = 77.008 \pm 1.985(6.626)$ which is (64, 90).

c) An increase of $100 is outside the confidence interval and so is a very implausible value for the population slope.

12.35 **Strength as leg press:**
a) i) <u>Assumptions</u>: Assume randomization, linear trend with normal conditional distribution for y and the same standard deviation at different values of x. These data were not gathered using randomization, and so inferences are tentative.
   ii) <u>Hypotheses</u>: The null hypothesis that the variables are independent is H_0: $\beta = 0$. The two-sided alternative hypothesis of dependence is H_a: $\beta \neq 0$.
   iii) <u>Test statistic</u>: According to software, the test statistic is 9.64. We also could calculate the test statistic as follows: $t = b/se = 5.2710/0.5469 = 9.64$.
   iv) <u>P-value</u>: From software, the P-value is 0.000.
   v) <u>Conclusion</u>: If H_0 were true that the population slope $\beta = 0$, it would be extremely unusual – the probability would be close to 0 – to get a sample slope at least as far from 0 as $b = 5.2710$. The P-value gives very strong evidence that an association exists between maximum leg press and number of 200-pound leg presses.

b) The 95% confidence interval is $b \pm t_{.025}(se) = 5.2710 \pm 2.004(0.5469)$

The confidence interval is (4.2, 6.4). The confidence interval, unlike the significance test, gives us a range of plausible values for the slope of the population.

12.37   More girls are good?:
   a)   The positive slope indicates a positive association between life length and number of daughters. Having more daughters is good.
   b)   i)   <u>Assumptions</u>: Assume randomization, linear trend with normal conditional distribution for y and the same standard deviation at different values of x.
        ii) <u>Hypotheses</u>: The null hypothesis that the variables are independent is H_0: $\beta = 0$. The two-sided alternative hypothesis of dependence is H_a: $\beta \neq 0$.
        iii) <u>Test statistic</u>: $t = b/se = 0.44/0.29 = 1.52$.
        iv) <u>P-value</u>: The P-value is 0.13..
        v)   <u>Conclusion</u>: If H_0 were true that the population slope $\beta = 0$, it would not be very unusual to get a sample slope at least as far from 0 as $b = 0.44$. The probability would be 0.13. It is plausible that there is no association between number of daughters and life length.
   c)   The 95% confidence interval is $b \pm t_{.025}(se) = 0.44 \pm 1.966(0.29)$, which is (-0.1, 1.0). The plausible values for the true population slope range from -0.1 to 1.0. Zero is a plausible value for this slope.

⌨12.39 Advertising and sales:
   a)   The mean for advertising is 2, and for sales is 7. The standard deviation for advertising is 2.16. The standard deviation for sales also is 2.16.

   b) $b = r\left(\dfrac{s_y}{s_x}\right) = 0.857(\dfrac{2.16}{2.16}) = 0.857$

        $a = \overline{y} - b\overline{x} = 7 - (0.857)(2) = 5.286$

        $\hat{y} = 5.286 + 0.857x$

   c)   i)   <u>Assumptions</u>: Assume randomization, linear trend with normal conditional distribution for y and the same standard deviation at different values of x.
        ii) <u>Hypotheses</u>: The null hypothesis that the variables are independent is H_0: $\beta = 0$. The two-sided alternative hypothesis of dependence is H_a: $\beta \neq 0$.
        iii) <u>Test statistic</u>: $t = b/se = 0.857/0.364 = 2.35$.
        iv) <u>P-value</u>: The P-value is 0.14.
        v)   <u>Conclusion</u>: If H_0 were true that the population slope $\beta = 0$, it would not be very unusual to get a sample slope at least as far from 0 as $b = 0.857$. The probability would be 0.14. The P-value is not below the significance level of 0.05, and, therefore, we cannot reject the null hypothesis. It is plausible that there is no association between advertising and sales.

⌨12.41 GPA and skipping class-revisited:
   $b \pm t_{.05}(se) = -.082 \pm -1.94(.016)$, which is (-.11, -.05). We are 90% confident that the population slope β falls between -.11 and -.05. On average, gpa decreases by between 0.11 to 0.05 points for every additional class that is skipped.

Section 12.4: Practicing the Basics

12.43   Poor predicted strengths:
   a)   The entry under BP_60, 15.0, is the number of 60 pound bench presses, and under BP, 105.00, is the maximum bench press for athlete 10. The predicted maximum bench press for this person, however, 85.90, is in the column "Fit." In the "Residual" column, we see the difference between the actual and predicted maximum bench press, 19.10. Finally, in the column titled "St Resid," we see the standardized residual, 2.4, which is the residual divided by the standard error that describes the sampling variability of the residuals; it does not depend on the units used to measure the variable. The "R" designates a standardized residual above 2.0.
   b)   We would expect about 5% of standardized residuals to have an absolute value above 2.0. Thus, it is not surprising that three would have an absolute value above 2.0.

12.45 Bench press residuals:
  a) This figure provides information about the distribution of standardized residuals, and hence the conditional distribution of maximum bench press.
  b) The distributions mentioned in (a) seem to be approximately normal.

12.47 Predicting clothes purchases:
  a) The value under "Fit," 448, is the predicted amount spent on clothes in the past year for those in the 12th grade of school.
  b) The 95% confidence interval of (427, 469) is the range of plausible values for the population mean of dollars spent on clothes for 12th grade students in the school.
  c) The 95% prediction interval of (101, 795) is the range of plausible values for the individual observations (dollars spent on clothes) for all the 12th grade students at the school.

12.49 ANOVA table for leg press:
  a) The residual standard deviation is the square root of the *MS Error*. The square root of 1303.72 is 36.1. This is the estimated standard deviation of maximum leg presses for female athletes who can do a fixed number of 200-pound leg presses.
  b) The standard deviation is 36.1; the 95% prediction interval for female athletes with $x = 22$ is $\hat{y} \pm 2s$ or $349.83 \pm 2(36.1)$, which is (277.6, 422.0).

12.51 Variability and *F*:
  a) The Total SS is the sum of the regression SS and the residual SS. The residual SS represents the error in using the regression line to predict y. The regression SS summarizes how much less error there is in predicting y using the regression line compared to using $\overline{y}$.
  b) The sum of squares around the mean divided by $n-1$ is $192,787/56 = 3442.6$, and its square root is 58.7. This estimates the overall standard deviation of y-values, whereas the residual s estimates the standard deviation of y-values at a fixed value of x.
  c) The F test statistic is 92.87; its square root is the t-statistic of 9.64.

12.53 Understanding an ANOVA table:
  a) The *MS* values will be calculated by dividing *SS* by *df*. The top *MS* will be $200,000/1 = 200,000$, and the bottom *MS* will be $700,000/31 = 22,580.6$. F is the ratio of the two mean squares, $200,000/22,580.6 = 8.86$.
  b) The F test statistic is an alternative test statistic for testing $H_0: \beta = 0$ against $H_a: \beta \neq 0$.

☐12.55 GPA ANOVA:
  Software gives us the ANOVA table:

```
Analysis of Variance
Source           DF      SS       MS       F       P
Regression        1    1.9577   1.9577   19.52   0.000
Residual Error   57    5.7158   0.1003
Total            58    7.6736
```

  a) The Total SS is the sum of the residual SS and regression SS. The Total SS = $7.6736 = 1.9577 + 5.7158$, where the residual SS = 5.7158 = the error in using the regression line to predict y. The regression SS = 1.9577 = how much less error there is in predicting y using the regression line compared to using $\overline{y}$.
  b) The estimated residual standard deviation of y is the square root of the *MS Error*; the square root of 0.1003 is 0.32. It estimates the standard deviation of y at fixed value of x, and describes typical size of the residuals about the regression line.
  c) The sample standard deviation, s_y, is the square root of the mean square total. *MS Total* = $7.6736/58 = 0.1323$. Its square root is 0.36. s_y refers to *all* the y-values, not just those at a particular x-value. It describes variability about the overall mean of $\overline{y}$ for those at all values of x, not just those at a specific value of x.

SECTION 12.5: PRACTICING THE BASICS

12.57 Growth by year vs. decade:
a) $(1.072)^{10} = 2.0$
b) $(1.10)^{10} = 2.59$. Your savings would double in a decade if you *added* 10% a year, but the effect here is multiplicative.

12.59 Future shock:
a) $1.15^5 = 2.0$; the population size after five decades is predicted to be 2.0 times the original population size.
b) $1.15^{10} = 4.0$; the population size after ten decades is predicted to be 4.0 times the original population size.
c) $1.15^{20} = 16.4$; the population size after 20 decades is predicted to be 16.4 times the original population size.

⊟12.61 Leaf litter decay:
a) A straight-line model is inappropriate since the scatterplot shows a curvilinear relationship.

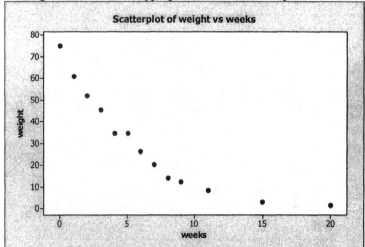

b) Software gives a straight-line regression equation of: `weight = 55.0 - 3.59 weeks`
   The predicted weight after 20 weeks is $55.0 - 3.59(20) = -16.8$; this does not make sense because a weight cannot be negative.
c) A straight-line model seems appropriate for the log of *y* against *x*.

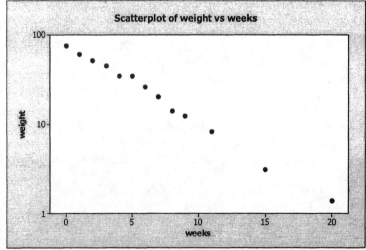

d) (i) $\hat{y} = 80.6(0.813)^0 = 80.6$; (ii) $\hat{y} = 80.6(0.813)^{20} = 1.3$
e) The coefficient 0.813 indicates that the predicted weight multiplies by 0.813 each week.

CHAPTER PROBLEMS: PRACTICING THE BASICS

12.63 **Parties and sports:**
   a) At a fixed values of x there is variability in the values of y so we can't specify individual y-values using x but we can try to specify the mean of those values and how that mean changes as x changes.
   b) Because y-values vary at a fixed value of x, the model has a σ parameter to describe the spread of the conditional distribution of those of those y-values at each fixed x.

12.65 **Short people:**
The response variable is height of children and the explanatory variable is height of parents. Very short parents tend to have children who are short, but not as short as they are. The prediction equation is based on correlation. Because the correlation between two variables is never greater than the absolute value of 1, a y value tends to be not so far from its mean as the x value is from its mean. This is called regression toward the mean.

12.67 **Bedroom residuals:**
   a) $\hat{y} = 33,778 + 31,077(3) = 127,009$; the residual is $338,000 - 127,009 = 210,991$. This house sold for $210,991 more than predicted.
   b) The standardized residual of 4.02 indicates that this observation is 4.02 standard errors higher than predicted.

12.69 **Types of spread:**
   a) The residual standard deviation of y refers to the spread of the y-values at a particular x-value, whereas the standard deviation refers to the spread of all of the y-values.
   b) The fact that they're not very different indicates that the number of bedrooms is not strongly associated with selling price. The spread of y-values at a given x is about the same as the spread of all of the y observations. We can see this by considering the r^2 of 0.13. The error using $\hat{y}$ to predict y is only 13% smaller than the error using $\overline{y}$ to predict y.

12.71 **Bench press predicting leg press:**
   a) The 95% confidence interval provides a range of plausible values for the population mean of y when $x = 80$. The plausible values range from 338 to 365 for the *mean* of y-values for *all* female high school athletes having $x = 80$.
   b) The prediction interval provides a range of predicted y-values for an individual observation when $x = 80$. For *all* female high school athletes with a maximum bench press of 80, we predict that 95% of them have maximum leg press between about 248 and 455 pounds. The 95% PI is for a single observation y, whereas the confidence interval is for the *mean* of y.

12.73 **Savings grow:**
   a) $1000 \times 2^5 = 32,000$
   b) $1000 \times 2^{10} = 1,024,000$
   c) The equation based on decade instead of year is: 1000×2^x

12.75 **World population growth:**
   a) 1900: $\hat{y} = 1.4193 \times 1.014^0 = 1.42$ billion;   2000: $\hat{y} = 1.4193 \times 1.014^{100} = 5.70$ billion
   b) The fit of the model corresponds to a rate of growth of 1.4% per year because multiplying by 1.014 adds an additional 1.4% each year.
   c) (i) The predicted population size doubles after 50 years because $1.014^{50} = 2.0$, the number by which we'd multiply the original population size.
      (ii) It quadruples after 100 years: $1.014^{100} = 4.0$
   d) The exponential regression model is more appropriate for these data because the log of the population size and the year number are more highly correlated than are the population size and the year number.

CHAPTER PROBLEMS: CONCEPTS AND INVESTIGATIONS

⌨**12.77 Softball data:**

a) The 3 outlying points represent outliers – values more than 1.5×IQR beyond either Q1 or Q3.

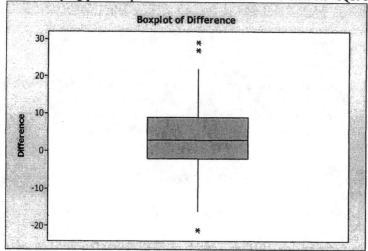

b) From software: Difference = - 9.125 + 1.178 Run
   Difference is positive when -9.125 + 1.178 (runs) > 0, which is equivalent to runs > 9.125/1.178 = 7.7; thus, the team scores more runs than their opponents with eight or more runs.

c) Runs, hits, and difference are positively associated with one another. Errors are negatively associated with those three variables.

	Run	Hits	Errors
Hits	0.819		
	0.000		
Errors	-0.259	-0.154	
	0.000	0.010	
Difference	0.818	0.657	-0.501
	0.000	0.000	0.000

d) From software:

Predictor	Coef	SE Coef	T	P
Constant	-9.1254	0.5934	-15.38	0.000
Run	1.17844	0.04989	23.62	0.000

The P-value is very small, and we would conclude that runs are positively associated with the difference scores.

⌨**12.79 GPA and TV watching:**

The two-page report will be different for each student, but would interpret results from the following output from software.

The regression equation is: high_sch_GPA = 3.44 - 0.0183 TV

Predictor	Coef	SE Coef	T	P
Constant	3.44135	0.08534	40.32	0.000
TV	-0.018305	0.008658	-2.11	0.039

S = 0.446707   R-Sq = 7.2%   R-Sq(adj) = 5.6%

Analysis of Variance

Source	DF	SS	MS	F	P
Regression	1	0.8921	0.8921	4.47	0.039
Residual Error	58	11.5737	0.1995		
Total	59	12.4658			

12.81 **Football point spreads**:
a) If there is no bias in the Las Vegas predictions, the predictions should exactly match the observations. In other words, y=x which translates to the true y-intercept equaling 0 and the true slope equaling 1.
b) No. Based on the results in the table, the P-value for testing that the true y-intercept is 0 is quite large so that we are unable to conclude that the y-intercept differs from 0. The P-value for testing that the slope is equal to 0 is approximately 0 so that we reject the null hypothesis of the true slope equaling 0. The least squares fit for the slope is quite close to 1, namely 1.0251.

12.83 **Sports and regression**:
There are many examples that students could use in their response. Here's one example. If a Major League baseball player has an amazing year with many homeruns, he's unlikely to have that many the following year. If we look at the top ten homerun hitters for a given year, most of them are likely to have fewer homeruns the following year. Once a hitter reaches a given level, he's more likely to come back toward the average; it's hard to go up at that point.

12.85 **Height and weight**:
As the range of values reflected by each sample is restricted, the correlation tends to decrease when we consider just students of a restricted range of ages. Using the two samples will increase the ranges for height and weight, and likely increase the correlation.

12.87 **Dollars and pounds**:
a) The slope would change because it depends on units. The slope would be 2 times the original slope.
b) The correlation would not change because it is independent of units.
c) The t-statistic would not change because although the slope doubles, so does its standard error. (The results of a test should not depend on the units we use.)

12.89 **df for t tests in regression**:
a) $df = n$ – the number of parameters; for the model, $\mu_y = \alpha + \beta x$, there are two parameters α and β, and so $df = n - 2$.
b) When the inference is about a single mean, there is only one parameter, and therefore, $df = n - 1$.

12.91 **Assumptions fail?**:
a) The percentage of unemployed workers would likely fluctuate quite a bit between 1900 and 2005, and this would not be a linear relationship.
b) Annual medical expenses would likely be quite high at low ages, then lower in the middle, then high again, forming a parabolic, rather than linear, relationship.
c) The relation between these variables is likely curvilinear. Life expectancy increases at first as per capita income increases, then gradually levels off.

12.93 **Growth in Florida**:
a) The statement is referring to additive growth, but this is multiplicative growth. There is an exponential relation between these variables.
b) $\hat{y} = 100{,}000 \times 1.042^{10} = 150{,}896$; the percentage growth for the decade is about 50.9%.

12.95 **Interpret r**:
The best response is (b).

12.97 **Slope and correlation**:
The best response is (d).

12.99 **Regress x on y**:
The best response is (a).

♦♦12.101 Golf club velocity and distance:
   a)   We would expect that at 0 impact velocity, there would be 0 putting distance. The line would pass through the point having coordinates (0, 0).
   b)   If x doubles, then x^2 (and hence the mean of y) quadruples. For example, if x goes from 2 to 4, then x^2 goes from 4 to 16.

♦♦12.103 r^2 and variances:

Because $r^2 = \dfrac{\sum(y-\bar{y})^2 - \sum(y-\hat{y})^2}{\sum(y-\bar{y})^2}$, it represents the relative difference between the quantity used to summarize the overall variability of the y values (i.e. the variability of the marginal distribution of y) and the quantity used to summarize the residual variability (i.e. the variance of the conditional distribution of y for a given x). These go in the numerator of the respective variance estimates, and their denominators are nearly identical ($n-1$ and $n-2$). Therefore, the estimated variance of the conditional distribution of y for a given x is approximately 30% smaller than the estimated variance of the marginal distribution of y.

♦♦12.105 Regression with an error term:
   a)   Error is calculated by subtracting the mean from the actual score, y. If this difference is positive, then the observation must fall above the mean.
   b) $\varepsilon = 0$ when the observation falls exactly at the mean. There is no error in this case.
   c)   Because the residual $e = y - \hat{y}$, we have $y = \hat{y} + e = a + bx + e$. As $\hat{y}$ is an estimate of the population mean, e is an estimate of ε.
   d)   It does not make sense to use the simpler model, $y = \alpha + \beta x$, that does not have an error term because it is improbable that every observation will fall exactly on the regression line; it is improbable that there will be no error.

CHAPTER PROBLEMS: STUDENT ACTIVITIES

⌨12.107 Analyze your data:
   Responses will be different based on the data files for each class and the variables chosen by each instructor.

Chapter 13
Multiple Regression

SECTION 13.1: PRACTICING THE BASICS

13.1   **Predicting weight**:
a) $\hat{y} = -121 + 3.50\,x_1 + 1.35\,x_2$
      $= -121 + 3.50\,(66) + 1.35\,(18) = 134.3$
b)   The residual $= y - \hat{y} = 115 - 134.3 = -19.3$. The actual total body weight is 19.3 pounds lower than predicted.

13.3   **Predicting college GPA**:
a)   (i) $\hat{y} = 0.20 + 0.50\,x_1 + 0.002\,x_2$
        $= 0.20 + 0.50(4.0) + 0.002(800) = 3.80$
   (ii) $\hat{y} = 0.20 + 0.50\,x_1 + 0.002\,x_2$
        $= 0.20 + 0.50(2.0) + 0.002(200) = 1.60$
b) $\hat{y} = 0.20 + 0.50\,x_1 + 0.002(500)$
     $= 0.20 + 0.50\,x_1 + 1 = 1.20 + 0.50\,x_1$
c) $\hat{y} = 0.20 + 0.50\,x_1 + 0.002(600)$
     $= 0.20 + 0.50\,x_1 + 1.2 = 1.40 + 0.50\,x_1$

13.5   **Does more education cause more crime?**:
a)   (i) $\hat{y} = 59.12 - 0.5834\,x_1 + 0.6825\,x_2$
        $= 59.12 - 0.5834(70) + 0.6825(0) = 18.3$
   (ii) $\hat{y} = 59.12 - 0.5834\,x_1 + 0.6825\,x_2$
        $= 59.12 - 0.5834(80) + 0.6825(0) = 12.4$
b)   When we control for urbanization, crime rate changes by the slope multiplied by the change in education. When education goes up $10 = 80 - 70$, predicted crime rate changes by ten multiplied by the slope, $10(-0.5834) = -5.8$
c)   (i) $\hat{y} = 59.12 - 0.5834\,x_1 + 0.6825(0) = 59.12 - 0.5834\,x_1$
   (ii) $\hat{y} = 59.12 - 0.5834\,x_1 + 0.6825(50) = 93.2 - 0.5834\,x_1$
   (iii) $\hat{y} = 59.12 - 0.5834\,x_1 + 0.6825(100) = 127.4 - 0.5834\,x_1$

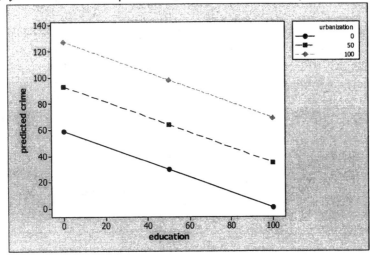

d) The line passing through the points having urbanization = 50 has a negative slope. The line passing through all the data points has a positive slope. Simpson's paradox occurs because the association between crime rate and education is positive overall but is negative at each fixed value of urbanization. It happens because urbanization is positively associated with crime rate and with education. As urbanization increases, so do crime rate and education tend to increase, giving an overall positive association between crime rate and education.

(i)

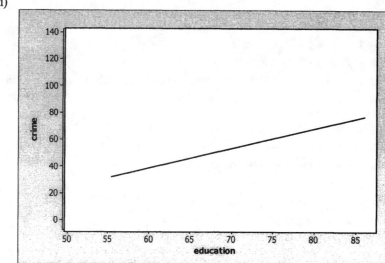

(ii)

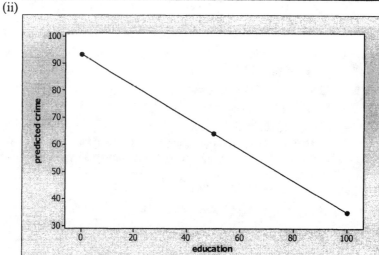

13.7   Putting or driving more important in golf?:

a) $\hat{y} = 261.78 - 0.0482\,x_1 + 0.0029\,x_2 - 1.307\,x_3 + 0.852\,x_4 = 261.78 - 0.0482(280) + 0.0029(40) - 1.307(40) + 0.852(110) = 289.8$

b) The coefficient for PUTTS is its slope. The predicted total score will increase by 0.85 for each increase of one in the number of putts taken, when controlling for the other variables in the model.

c) (i)   change of -0.48
    (ii) change of 0.03
    (iii) change of -13.07
    (iv) change of -8.52
    Increasing GIR by 10 led to the largest predicted decrease (see (iii))

13.9 **Controlling can have no effect**:
If x_1 and x_2 are not related to each other, then their slopes would be the same as if each were in a bivariate regression equation by itself. We don't need to control x_2 if it's not related to x_1. Changes in x_2 will not have an impact on the effect of x_1 on y.

Section 13.2: Practicing the Basics

13.11 **Predicting golf scores**:
a) $R^2 = \dfrac{\Sigma(y-\overline{y})^2 - \Sigma(y-\hat{y})^2}{\Sigma(y-\overline{y})^2} = (1886-287)/1886 = 0.848$

b) Using these variables together to predict total score reduces the prediction error by 85%, relative to using $\overline{y}$ alone to predict total score. Yes, the prediction is much better.

c) $R = \sqrt{R^2} = \sqrt{0.848} = 0.92$; there is a strong association between the observed total score and the predicted total score.

13.13 **When does controlling have little effect?**:
Controlling for body fat and then age does not change the effect of height much because height is not strongly correlated with either body fat or age.

13.15 **More Internet use**:
Because x_1 and x_2 are themselves highly correlated, once one of them is in the model, the remaining one does not help much in adding to the predictive power.

13.17 **Slopes, correlations, and units**:
a) The correlation between predicted house selling price and actual house selling price is 0.84
b) If selling price is measured in thousands of dollars, each y-value would be divided by 1,000. For example, 145,000 dollars would become 145 thousands of dollars. Each slope also would be divided by 1,000 (e.g., a slope of 53.8 for house size on selling price in dollars corresponds to 0.0538 in thousands of dollars for $\hat{y}$).
c) The multiple correlation would not change because it is not dependent on units.

Section 13.3: Practicing the Basics

13.19 **Predicting GPA**:
a) If the population slope coefficient equals 0, it means that, in the population of all students, high school GPA doesn't predict college GPA for students having any given value for study time. For example, for students who study 5 hours, high school GPA does not predict college GPA.
b) 1) <u>Assumptions</u>: We assume a random sample and that the model holds (each explanatory variable has a straight-line relation with μ_y, controlling for the other predictors, with the same slope for all combinations of values of other predictors in model, and there is a normal distribution for y with the same standard deviation at each combination of values of the other predictors in the model). Here, the 59 students were a convenience sample, not a random sample, so inferences are highly tentative.

    2) <u>Hypotheses</u>: H_0: $\beta_1 = 0$; H_a: $\beta_1 \neq 0$

    3) <u>Test statistic</u>: $t = (b_1 - 0)/se = 0.6434/0.1458 = 4.41$

    4) <u>P-value</u>: 0.000

    5) <u>Conclusion</u>: The P-value of 0.000 gives evidence against the null hypothesis that $\beta_1 = 0$. If the null hypothesis were true, the probability would be almost 0 of getting a test statistic at least as extreme as the value observed. We have very strong evidence that high school GPA predicts college GPA, if we already know study time. At common significance levels, such as 0.05, we reject H_0.

13.21 **Variability in college GPA**:
a) The residual standard deviation, 0.32, describes the typical size of the residuals and also estimates the standard deviation of *y* at fixed values of the predictors. For students with certain fixed values of high school GPA and study time, college GPAs vary with a standard deviation of 0.32.
b) Approximately 95% of college GPAs fall within about 2*s*, 0.64, of the true regression equation. When high school GPA = 3.80 and study time = 5.0, college GPA is predicted to be 3.61. Thus, we would expect that approximately 95% of the Georgia college students fall between 2.97 and 4.25.

13.23 **Leg press uncorrelated with strength?**:
The first test analyzes the effect of LP_200 at any given fixed value of BP_60, whereas the second test describes the overall effect of LP_200 ignoring other variables. These are different effects, so one can exist when the other does not. In this case, it is likely that LP_200 and BP_60 are strongly associated with one another, and the effect of LP_200 is weaker once we control for BP_60.

13.25 **Any predictive power?**:
a) H_0: $\beta_1 = \beta_2 = 0$; the null hypothesis states that neither of the two explanatory variables has an effect on the response variable *y*?
b) 3.16
c) The observed *F* statistic is 51.39 with a P-value of 0.000. If the null hypothesis were true, the probability would be close to 0 of getting a test statistic at least as extreme as the value observed. This P-value gives extremely strong evidence against the null hypothesis that $\beta_1 = \beta_2 = 0$. At common significance levels, such as 0.05, we can reject H_0. At least one of the two explanatory variables has an effect on BP.

13.27 **Regression for mental health**:
a) $b_1 \pm t_{.025} (se) = 0.10326 \pm 2.026(0.03250) = 0.10326 \pm 0.06585$, or (0.04, 0.17).
b) The confidence interval gives plausible values for the slope, the amount that mean mental impairment will increase when life events score increases by one, when controlling for SES. For an increase in 100 units of life events, we multiply each endpoint of the confidence interval by 100, giving (4, 17); this indicates that plausible increases in mean mental impairment range from 4 to 17 when life events scores increase by 100, controlling for SES.

13.29 **More predictors for house price**:
a) H_0: $\beta_1 = \beta_2 = \beta_3 = 0$ means that house selling price is independent of size of home, lot size, and real estate tax.
b) The large *F* value and small P-value provide strong evidence that at least one of the three explanatory variables has an effect on selling price.
c) The results of the *t* tests tell us that each of the three explanatory variables contributes to the prediction of selling price. All three are statistically significant predictors when controlling for the other explanatory variables.

Section 13.4: Practicing the Basics

13.31 **Body weight residuals**:
a) These give us information about the conditional distribution.
b) The histogram suggests that the distribution may be skewed to the right, rather than normal.

13.33 **More residuals for strength**:
One might think that this suggests less variability at low levels and even less at high levels of BP_60, but this may merely reflect fewer points in those regions. Overall, it seems OK.

13.35 **Driving accidents:**
  a) Example of possible scatterplot:

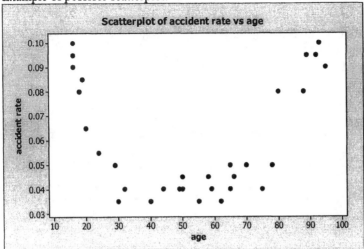

  b) Example of plot of standardized residuals against the values of age:

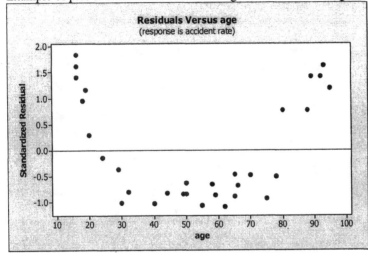

13.37 **College athletes**:
  a) The upper right and middle right plots both show strong, positive associations with BP.
  b) $\hat{y} = 55.012 + 0.16676LBM + 1.6575REP\_BP$; 1.66 is the amount that predicted maximum bench press changes for a one unit increase in number of repetitions, controlling for lean body mass.
  c) R^2 is 0.832. Using these variables together to predict BP reduces the prediction error by 83%, relative to using $\overline{y}$ alone to predict BP.
  d) The multiple correlation is 0.91. There is a strong association between the observed and predicted BPs.
  e) $F = (7641.5/50.7) = 150.75$; P-value is 0.000. If the null hypothesis were true, the probability would be close to 0 of getting a test statistic at least as extreme as the value observed. We have very strong evidence that BP is not independent of these two predictors.

f) 1) Assumptions: We assume a random sample and that the model holds (each explanatory variable has a straight-line relation with μ_y, with same slope for all combinations of values of other predictors in model, and there is a normal distribution for y with the same standard deviation at each combination of values of the other predictors in the model). Here, the 64 athletes were a convenience sample, not a random sample, so inferences are tentative.

2) Hypotheses: H_0: $\beta_1 = 0$; H_a: $\beta_1 \neq 0$

3) Test statistic: $t = 2.22$

4) P-value: 0.030

5) Conclusion: If the null hypothesis were true, the probability would be 0.03 of getting a test statistic at least as extreme as the value observed. The P-value gives relatively strong evidence against the null hypothesis that $\beta_1 = 0$.

g) The histogram suggests that the residuals are roughly bell-shaped about 0. They fall between about -3 and +3. The shape suggests that the conditional distribution of the response variable is roughly normal.

h) The plot of residuals against values of REP_BP describes the degree to which the response variable is linearly related to this particular explanatory variable. It suggests that the residuals are less variable at smaller values of REP_BP than at larger values of REP_BP.

i) The individual with REP_BP around 32 and standardized residual around -3 had a BP value considerably lower than predicted.

13.39 **Selling prices level off**:
Example of possible plot:

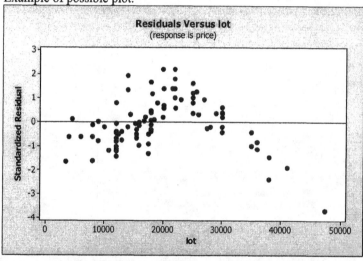

Section 13.5: Practicing the Basics

13.41 **eBay selling prices**:
a) For each increase in one bid, predicted selling price decreases by 0.72 for the same buying option.
b) When the buy-it-now option was used, predicted selling price was 2.77 lower.

13.43 **Quality and productivity:**

a) (i) Minimum: 61.3+0.35(12)=65.5; (ii) Maximum: 61.2+0.35(54)=80.2

b) When controlling for region, an increase of one hour leads to a decrease in predicted defects of 0.78 per 100 cars. Japanese facilities had 36 fewer predicted defects, on average, than did other facilities.

c) Simpson's paradox has occurred because the direction of the association between time and defects reversed when the variable of whether facility is Japanese was added.

d) Simpson's paradox occurred because, overall, Japanese facilities have fewer defects and take less time, whereas other facilities have more defects and take more time. When the data are looked at together, this leads to an overall positive association between defects and time.

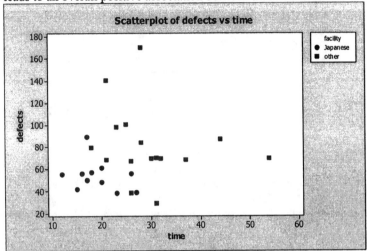

13.45 Houses, tax, and NW:

a) From software: `price = 43014 + 45.3 Taxes + 10814 NW`
   For homes in the NW, price = 53,828 + 45.3Taxes. For homes not in the NW, price = 43,014 + 45.3Taxes.

b) The coefficient, 10814, indicates that the predicted selling price for houses in the Northwest is 10,814 higher than the predicted selling price for houses elsewhere.

13.47 **Equal high jump for men and women:**

a) interaction

b) Males: $\hat{y} = 1.94 + 0.0064(875) = 7.54$

   Females: $\hat{y} = 1.59 + 0.0068(875) = 7.54$

c) It is not sensible to use this model to predict for the year 2803 because it is dangerous to extrapolate beyond our existing data. We do not know whether trends will change in the future.

Section 13.6: Practicing the Basics

13.49 **Income and credit cards:**

$$\hat{p} = \frac{e^{-3.52+0.105x}}{1+e^{-3.52+0.105x}} = \frac{e^{-3.52+0.105(25)}}{1+e^{-3.52+0.105(25)}} = \frac{e^{-0.895}}{1+e^{-0.895}} = (0.41/1.41) = 0.29$$

13.51 **Horseshoe crabs:**

a) Q1: $\dfrac{e^{-3.695+1.815x}}{1+e^{-3.695+1.815x}} = \dfrac{e^{-3.695+1.815(2.00)}}{1+e^{-3.695+1.815(2.00)}} = 0.937/1.937 = 0.48$

   Q3: $\dfrac{e^{-3.695+1.815(2.85)}}{1+e^{-3.695+1.815(2.85)}} = 4.383/5.383 = 0.81$

b) $0.81 - 0.48 = 0.33$. The probability increases by 0.33 over the middle half of the sampled weights.

13.53 **Voting and income**:

a) $\dfrac{e^{-1.00+0.02(10)}}{1+e^{-1.00+0.02(10)}} = 0.4493/1.4493 = 0.31$

b) $\dfrac{e^{-1.00+0.02(100)}}{1+e^{-1.00+0.02(100)}} = 2.718/3.718 = 0.73$

The predicted probability of voting Republican increases quite a bit as income increases.

13.55 **Many predictors of voting**:

a) As family income increases, people are, on average, more likely to vote Republican. As number of years of education increases, people are, on average, more likely to vote Republican. Men are more likely, on average, to vote Republican than are women.

b) (a) $\hat{p} = \dfrac{e^{-2.40+0.02x_1+0.08x_2+0.20x_3}}{1+e^{-2.40+0.02x_1+0.08x_2+0.20x_3}} = \dfrac{e^{-2.40+0.02(40)+0.08(16)+0.20(1)}}{1+e^{-2.40+0.02(40)+0.08(16)+0.20(1)}} = \dfrac{e^{-0.12}}{1+e^{-0.12}} = 0.8869/1.8869 = 0.47$

(b) $\hat{p} = \dfrac{e^{-2.40+0.02(40)+0.08(16)+0.20(0)}}{1+e^{-2.40+0.02(40)+0.08(16)+0.20(0)}} = \dfrac{e^{-0.32}}{1+e^{-0.32}} = 0.7261/1.7261 = 0.42$

13.57 **Death penalty and race**:

a) Controlling for victim's race, the proportion of black defendants who received the death penalty was $15/191 = 0.079$, and the proportion of white defendants who received the death penalty was $53/483 = 0.110$.

b) According to this equation, the death penalty is predicted to be most likely for black defendants who had white victims. We know this since the coefficient for defendant race is negative; thus 0 (black) leads to a higher predicted death penalty proportion than 1 (white). Also, the coefficient for victim's race is positive; therefore, 1 (white) would lead to a higher predicted death penalty proportion than would 0 (black).

CHAPTER PROBLEMS: PRACTICING THE BASICS

13.59 House prices:

a)

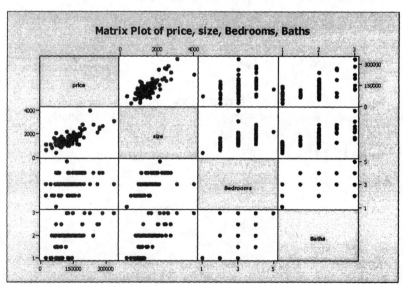

The plots that pertain to selling price as a response variable are those across the top row. The highly discrete nature of x_2 and x_3 limits the number of values these variables can take on. This is reflected in the plots, particularly the plot for bedrooms by baths.

b) From software: `price = 13596 + 74.1 size - 11164 Bedrooms + 17651 Baths`
   When number of bedrooms and number of bathrooms are fixed, an increase of one in size of home leads to an increase of 74.1 in predicted selling price.

c) $R^2 = [(3.14 \times 10^{11}) - (1.25 \times 10^{11})] / (3.14 \times 10^{11}) = 0.602$; this indicates that predictions are 60% better when using the prediction equation instead of using the sample mean $\overline{y}$ to predict y.

d) The multiple correlation, 0.78, is the square root of R^2. It is the correlation between the observed y-values and the predicted $\hat{y}$-values.

e) (i) Assumptions: multiple regression equation holds, data gathered randomly, normal distribution for y with same standard deviation at each combination of predictors.

   (ii) Hypotheses: H_0: $\beta_1 = \beta_2 = \beta_3 = 0$; H_0: At least one β parameter differs from 0.

   (iii) Test statistic: $F = 63,148,663,440/1,301,943,013 = 48.5$

   (iv) P-value: for df (3,96); 0.000

   (v) Conclusion: If the null hypothesis were true, the probability would be close to 0 of getting a test statistic at least as extreme as the value observed. We have very strong evidence that at least one explanatory variable has an effect on y.

f) The t statistic is -1.64 with a one-sided P-value of 0.103/2 = 0.052. If the null hypothesis were true, the probability would be 0.052 of getting a test statistic at least as extreme as the value observed. At a significance level of 0.05, we cannot reject the null. It is plausible that the number of bedrooms does not have an effect on selling price. This is likely not significant because it is correlated with the other explanatory variables in this model. It might be associated with selling price on its own, but might not provide additional predictive information over and above the other explanatory variables.

g)

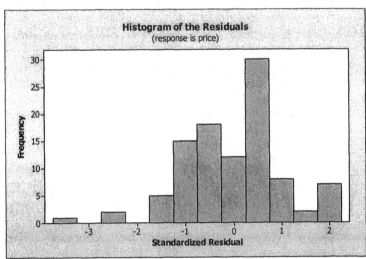

This histogram describes the shape of the conditional distribution of y at given values of the explanatory variables. It suggests that the distribution may be skewed to the left.

h)

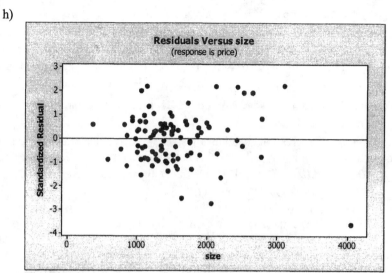

This plot depicts the size of the residuals for the different house sizes observed in this sample. It indicates possibly greater residual variability (and hence, greater variability in selling price) as house size increases.

■13.61 Softball data:

a) From software, the prediction equation is: `Difference = - 5.00 + 0.934 Hits - 1.61 Errors;` the slopes indicate that for each increase of one hit, the predicted difference increases by 0.93, and for each increase of one error, the predicted difference decreases by 1.61.

b) When errors = 0, the prediction equation is $\hat{y} = -5.00 + 0.934\ x_1$; for a predicted difference of 0, we would need 5.35 hits. ($0 = -5.00 + 0.934\ x_1$; $x_1 = 5.35$). Thus, the team would need six or more hits so that the predicted difference is positive (if they can play error-free ball).

13.63   Effect of poverty on crime:

The slope with x_3 in the model represents the effect of poverty when controlling for percentage of single-parent families, as well as percent living in urban areas. The slope without x_3 in the model represents the effect of poverty when controlling only for percent living in urban areas.

13.65   Significant fertility prediction?:

a) $F = 20.742/0.484 = 42.9$; P-value: 0.000; if the null hypothesis were true, the probability would be close to 0 of getting a test statistic at least as extreme as the value observed. We have very strong evidence that at least one of these explanatory variables predicts y better than the sample mean does.

b) The significance test would not be relevant if we were not interested in nations beyond those in the study; if this were the case, the group of nations that we studied would be a population and not a sample.

13.67   Education and gender in modeling income:

Since the effect of education on income changes depending on gender, the explanatory variables, education and gender, are said to interact.

13.69   AIDS and AZT:

a) The negative sign indicates that if an individual used AZT, the predicted probability of developing AIDS symptoms was lower.

b) black/yes: $\hat{p} = \dfrac{e^{-1.074-0.720(1)+0.056x(0)}}{1+e^{-1.074-0.720(1)+0.056(0)}} = 0.1663/1.1663 = 0.14$

black/no: $\hat{p} = \dfrac{e^{-1.074-0.720(0)+0.056x(0)}}{1+e^{-1.074-0.720(0)+0.056(0)}} = 0.3416/1.3416 = 0.26$

c) 1) Assumptions: The data were generated randomly. The response variable is binary.

2) Hypotheses: H_0: $\beta_1 = 0$; H_a: $\beta_1 \neq 0$

3) Test statistic: $z = (b-0)/se = (-.720 - 0)/0.279 = -2.58$

4) P-value: 0.010

5) Conclusion: If the null hypothesis were true, the probability would be 0.01 of getting a test statistic at least as extreme as the value observed. We have very strong evidence against the null hypothesis that $\beta_1 = 0$. At significance level of 0.05, we can reject H_0.

CHAPTER PROBLEMS: CONCEPTS AND INVESTIGATIONS

🖳13.71 Student data:

The reports will be different for each student, but could include the following, along with associated graphs.

```
The regression equation is
college_GPA = 2.83 + 0.203 high_sch_GPA - 0.0092 sports
Predictor        Coef  SE Coef      T      P
Constant       2.8293   0.3385   8.36  0.000
high_sch_GPA   0.20309  0.09753  2.08  0.042
sports        -0.00922  0.01161 -0.79  0.430

S = 0.341590   R-Sq = 8.8%   R-Sq(adj) = 5.6%

Analysis of Variance
Source         DF      SS      MS     F      P
Regression      2  0.6384  0.3192  2.74  0.073
Residual Error 57  6.6510  0.1167
Total          59  7.2893

Unusual Observations
Obs  high_sch_GPA  college_GPA     Fit  SE Fit  Residual  St Resid
 11          2.30       2.6000  3.1580  0.1476   -0.5580  -1.81 X
 42          2.00       3.0000  3.1616  0.1351   -0.1616  -0.52 X
 50          3.00       4.0000  3.3002  0.1224    0.6998   2.19R
 60          3.40       3.0000  3.3722  0.1345   -0.3722  -1.19 X
```

R denotes an observation with a large standardized residual.
X denotes an observation whose X value gives it large influence.

13.73   Modeling salaries:

One-page reports will be different for each student, but should indicate that the three non-categorical explanatory variables – years of experience, degree quality, and research productivity – are all positively associated with salaries, given the other explanatory variables in the model. For the categorical variables, predicted salary is higher among those with a Ph.D. as a terminal degree, as well as among those who are married, white, or male.

13.75   Interpret indicator:

The best answer is (d).

13.77 **True or false about R and R^2:**
a) False; the multiple correlation is at least as large as the ordinary correlations, but could be larger; it is the correlation between the observed and predicted value of y.
b) False; it falls between 0 and 1.
c) False; R^2 describes how well you can predict y using a set of explanatory variables together in a multiple regression model
d) True

13.79 **True or false about slopes:**
a) True; the slope for this variable is positive in the bivariate regression equation.
b) False; a one-unit increase in x_1 corresponds to a change of 0.45 in the predicted value of y, only when we ignore x_2.
c) True; the slope for x_2 is 0.003 when controlling for x_1. 0.003 multiplied by 100 is 0.30.

13.81 **Lurking variable:**
y = math achievement score, x_1 = height, x_2 = age for a sample of children from all the different grades in a school system.

13.83 **Why an F test?:**
When doing multiple significance tests, one may be significant merely by random variation. When there are many explanatory variables, doing the F test first provides protection from doing lots of t tests and having one of them be significant merely by random variation when, in fact, there truly are no effects in the population.

13.85 **Logistic vs. linear:**
When $x=0$, a little extra income is not going to make a difference; one likely can't afford a home no matter what. Similarly, when $x=50,000$, a little extra income won't make much difference; one likely can afford a home no matter what. Only in the middle is the extra income likely to "push" someone over the income level at which he or she can afford a home. In such a case, a linear regression model would not be appropriate, although a logistic regression model would.

♦♦13.87 R can't go down:
When you add a predictor, if it has no effect its coefficient is 0. Then the prediction equation is exactly the same as with the simpler model without that variable and R will be exactly the same as before. If having a nonzero coefficient results in better predictions overall, then R will increase.

♦♦13.89 Simpson's paradox:

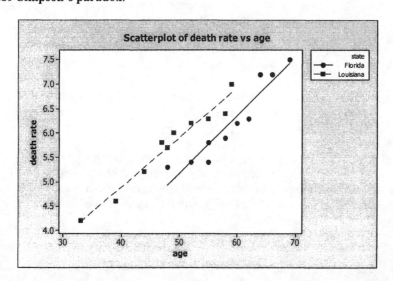

◆◆**13.91 Logistic slope**:
a) When $p = 0.5$, $p(1-p) = 0.5(1-0.5) = 0.25$; 0.25 multiplied by β is the same as β/4.
b) $0.1(1-0.1) = 0.09$; $0.3(1-0.3) = 0.21$; $0.7(1-0.7) = 0.21$; $0.9(1-0.9) = 0.09$; as p gets closer and closer to 1, the slope approaches 0.

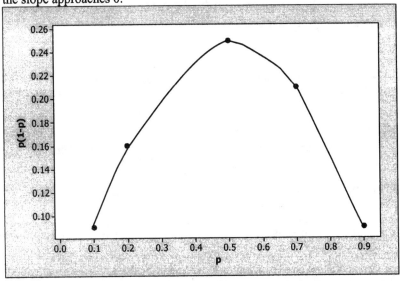

CHAPTER PROBLEMS: STUDENT ACTIVITIES

⌨**13.93 Class data**:
The responses will be different for each class.

Chapter 14
Comparing Groups: Analysis of Variance

SECTION 14.1: PRACTICING THE BASICS

14.1   **Hotel satisfaction:**
   a) The response variable is the performance gap, the factor is which hotel the guest stayed in, and the categories are the five hotels.
   b) H_0: $\mu_1 = \mu_2 = \mu_3 = \mu_4 = \mu_5$
   H_a: at least two of the population means are unequal.
   c) $df_1 = 4$ because there are five groups and $df_1 = g - 1$; $df_2 = 120$ because there are 125 people in the study and five groups, and $df_2 = N - g$.
   d) From a table or software, $F = 2.45$ and higher

14.3   **What's the best way to learn French?:**
   a)   i)   <u>Assumptions</u>: Independent random samples, normal population distributions with equal standard deviations
      ii)   <u>Hypotheses</u>: H_0: $\mu_1 = \mu_2 = \mu_3$
         H_a: at least two of the population means are unequal
      iii) <u>Test statistic</u>: $F = 2.50$ ($df_1 = 2$, $df_2 = 5$)
      iv)   <u>P-value</u>: 0.18
      v)   <u>Conclusion</u>: If the null hypothesis were true, the probability would be 0.18 of getting a test statistic at least as extreme as the value observed. There is not much evidence against the null. It is plausible that the null hypothesis is correct and that there is no difference among the population mean quiz scores of the three types of students.
   b) The P-value is not small likely because of the small sample sizes. Because the numerator of the between-groups variance estimate involves multiplying by the sample size n for each group, a smaller n leads to a smaller overall between-groups estimate of variance and a smaller test statistic value.
   c) This was an observational study because students were not assigned randomly to groups. There could have been a lurking variable, such as school GPA, that was associated with students' group membership, but also with the response variable. Perhaps higher GPA students are more likely to have previously studied a language and higher GPA students also tend to do better on quizzes than other students.

14.5   **Outsourcing:**
   a) H_0: $\mu_1 = \mu_2 = \mu_3$

      μ_1 represents the population mean satisfaction rating for San Jose, μ_2 for Toronto, and μ_3 for Bangalore.
   b) The F statistic, 27.6, can be calculated by dividing the between-groups variance estimate, 13.00, by the within-groups variance estimate, 0.47. The degrees of freedom are: $df_1 = 2$ and $df_2 = 297$.
   c) If the null hypothesis were true, the probability would be close to 0 of test statistic at least as extreme as the value observed. We have very strong evidence that customer satisfaction ratings are different for at least two of the populations. With a 0.05 significance level, we would reject the null hypothesis.

14.7 How many kids to have?:

a) μ_1 represents the population mean ideal number of kids for Protestants; μ_2 represents the population mean for Catholics; μ_3 for Jewish people; μ_4 for those of another religion; and μ_5 for those of no religion; H$_0$: $\mu_1 = \mu_2 = \mu_3 = \mu_4 = \mu_5$

b) The assumptions are that there are independent random samples, and normal population distributions with equal standard deviations.

c) The *F* statistic is 5.48, and the P-value is 0.001. If the null hypothesis were true, the probability would be 0.001 of getting a test statistic at least as extreme as the value observed. We have strong evidence that a difference exists between at least two of the population means for ideal numbers of kids.

d) We cannot conclude that every pair of religious affiliations has different population means. ANOVA tests only whether at least two population means are different.

▢14.9 Florida student data:

a)

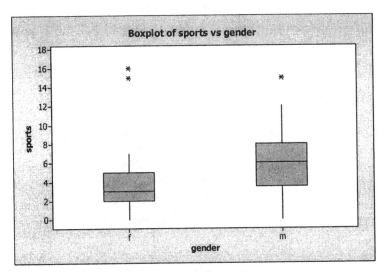

Variable	gender	N	Mean	StDev
sports	f	31	4.129	3.640
	m	29	6.310	3.828

b) Hypotheses: H$_0$: $\mu_1 = \mu_2$; H$_a$: at least two of the population means are unequal (of course, in this case, there are only two population means, so we could write $\mu_1 \neq \mu_2$). $F = 5.12$; P-value = 0.027. If the null hypothesis were true, the probability would be 0.027 of getting a test statistic at least as extreme as the value observed. We have strong evidence that a difference exists among the population mean number of weekly hours engaged in sports and other physical exercise.

c) The *t* statistic would be the square root of the *F* statistic, and the P-values would be identical.

14.11 Comparing therapies for anorexia:

a)

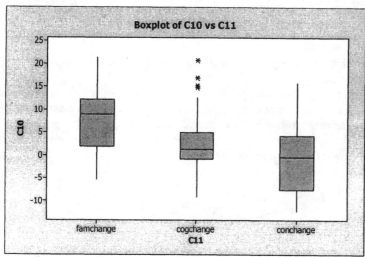

Level	N	Mean	StDev
famchange	17	7.265	7.157
cogchange	29	3.007	7.309
conchange	26	-0.450	7.989

The box plots and descriptive statistics suggest that the means of these groups are somewhat different. The standard deviations are similar.

b) $F = 5.42$; P-value = 0.006; if the null hypothesis were true, the probability would be 0.006 of getting a test statistic at least as extreme as the value observed. We have strong evidence that at least two population means are different.

c) The assumptions are that there are independent random samples, and normal population distributions with equal standard deviations. The assumptions are that there are independent random samples and normal population distributions with equal standard deviations. There is evidence of skew, but the test is robust with respect to this assumption. The subjects were randomly assigned to treatments but are not a random sample of subjects suffering from anorexia, so results are highly tentative.

SECTION 14.2: PRACTICING THE BASICS

14.13 Religiosity and number of good friends:

$$(\overline{y}_1 - \overline{y}_2) \pm t_{.025}\, s \sqrt{\frac{1}{n_1} + \frac{1}{n_2}} = (12.1 - 6.2) \pm 1.963(15.437)\sqrt{\frac{1}{337} + \frac{1}{330}} = 5.9 \pm 2.35; \text{ which is } (3.55, 8.25).$$

Because 0 does not fall in this confidence interval, we can infer at the 95% confidence level that the population means are different (higher for the high religiosity group than for the low religiosity group).

14.15 Tukey holding time comparisons:

a) The only significant difference now is that between classical music and Muzak.

b) The margins of error are larger than with the separate 95% intervals because the Tukey method uses an overall confidence level of 95% for the entire set of intervals.

14.17 REM regression:

a) $x_1 = 1$ for observations from the first group and $= 0$ otherwise; $x_2 = 1$ for observations from the second group and $= 0$ otherwise.

b) H_0: $\mu_1 = \mu_2 = \mu_3$; H_0: $\beta_1 = \beta_2 = 0$

c) The intercept estimate is the mean for the third group, which is 12. The first estimated regression coefficient is the difference between the means for the first and third groups, or $18 - 12 = 6$, and the second is the difference between the means for the second and third groups, or $15 - 12 = 3$.

14.19 **Regression for outsourcing**:

a) $x_1 = 1$ for observations from San Jose and $= 0$ otherwise; $x_2 = 1$ for observations from Toronto and $= 0$ otherwise.

b) 7.1 is the sample mean for Bangalore, 0.5 is the difference between the sample means for San Jose and Bangalore, and 0.7 is the difference between the sample means for Toronto and Bangalore.

14.21 French ANOVA:

a) Group 2 – Group 1: (-8.7, 2.7)
Group 3 – Group 1: (-3.1, 7.1)
Group 3 – Group 2: (-0.7, 10.7)
Because 0 falls in all three confidence intervals, we cannot infer that any of the pairs of population means are different.

b) Note: Statistical software such as MINITAB is needed to complete this solution.
Group 2 – Group 1: (-10.3, 4.3)
Group 3 – Group 1: (-4.5, 8.5)
Group 3 – Group 2: (-2.3, 12.3)
Again, 0 falls in all three confidence intervals; we cannot infer that any of the pairs of population means are different. The intervals are wider than those in (a) because we are now using a 95% confidence level for the overall set of intervals rather than for each separate interval.

SECTION 14.3: PRACTICING THE BASICS

14.23 **Drug main effects:** Hypothetical sets of population means will be different for each student. Means provided are examples of possible answers.

a)

	Lipitor	Pravachol
Low	10	10
High	20	20

b)

	Lipitor	Pravachol
Low	10	20
High	10	20

c)

	Lipitor	Pravachol
Low	10	20
High	20	30

d)

	Lipitor	Pravachol
Low	10	10
High	10	10

14.25 **House prices, region, and bedrooms**:

a) $F = 9128/2542 = 3.59$

b) The small P-value of 0.017 provides strong evidence that the population mean house selling price depends on region of the city. If the null hypothesis were true, the probability would be 0.017 of getting a test statistic at least as extreme as the value observed.

14.27 **Hang up if message repeated?**:

a) H_0: Population mean holding time is equal for the three types of messages, for each fixed level of repeat time.

b) $F = 74.60/10.52 = 7.09$; the small P-value of 0.011 provides strong evidence that the population mean holding time depends on the type of message. If the null hypothesis were true, the probability would be 0.011 of getting a test statistic at least as extreme as the value observed.

c) The assumptions for two-way ANOVA are that the population distribution for each group is normal, the population standard deviations are identical, and the data result from a random sample or randomized experiment.

14.29 Interaction between message and repeat time?:
a) H_0: no interaction; $F = 0.67$; P-value = 0.535 (rounds to 0.54)
b) If the null hypothesis were true, the probability would be 0.54 of getting a test statistic at least as extreme as the value observed. It is plausible that the null hypothesis is correct and that there is no interaction. This lends validity to the previous analyses that assumed a lack of interaction.

14.31 Ideology by gender and race:
a) The means for women are almost the same (black: 4.06; white: 4.04), whereas the mean for white males is about 0.5 higher than the mean for black males (black: 3.74; white: 4.25).
b) The gender effect was such that women had the higher mean for blacks but men had the higher mean for whites. Overall, ignoring race, the means may be quite similar for men and for women, and the one-way ANOVA for testing the gender effect may not be significant.
c) From a two-way ANOVA, we learn that the effect of gender differs based on race (as described above), but we do not learn this from the one-way ANOVA.

⌨14.33 Diet and weight gain:
a) $F = 13.90$; P-value = 0.000; the small P-value provides very strong evidence that the mean weight gain depends on the protein level.
b) H_0: no interaction; $F = 2.75$; P-value: 0.07; the P-value is not smaller than 0.05, so we cannot reject the null hypothesis. It is plausible that there is no interaction.
c) $se = s\sqrt{\dfrac{1}{n \text{ for low protein beef}} + \dfrac{1}{n \text{ for high protein beef}}} = 14.648\sqrt{\dfrac{1}{10} + \dfrac{1}{10}} = 6.551$

$(\overline{y}_i - \overline{y}_j) \pm t_{.025}\, se$; $(100.00 - 79.20) \pm 2.005(6.551)$, which is (7.7, 33.9)

CHAPTER PROBLEMS: PRACTICING THE BASICS

14.35 Good friends and marital status:
a) We can denote the number of good friends means for the population that these five samples represent by μ_1 for married, μ_2 for widowed, μ_3 for divorced, μ_4 for separated, and μ_5 for never married. The null hypothesis is H_0: $\mu_1 = \mu_2 = \mu_3 = \mu_4 = \mu_5$. The alternative hypothesis is that at least two of the population means are different.
b) No; large values of F contradict the null, and when the null is true the expected value of the F statistic is approximately 1.
c) If the null hypothesis were true, the probability would be 0.53 of getting a test statistic at least as extreme as the value observed. It is plausible that the null hypothesis is correct and that no difference exists among the five marital status groups in the population mean number of good friends.

14.37 TV watching:
a) $F = 249.93/6.41 = 39.0$; this test statistic has a P-value far smaller than 0.05; we can reject the null hypothesis, and can conclude that a difference exists among the three races in the population mean hours per day of TV watching.
b) Because 0 is not in the confidence intervals for the mean differences between blacks and whites and between blacks and other, we can conclude that these differences are statistically significant. It appears that blacks watch more hours of TV per day than either whites or those in the category "other."
c) $(4.14 - 2.75) \pm 1.961(2.53)\sqrt{\dfrac{1}{305} + \dfrac{1}{1435}} = 1.39 \pm 0.31$, which is (1.1, 1.7).
d) The corresponding interval formed with the Tukey method would be wider because it uses a 95% confidence level for the overall set of intervals, rather than for each one separately.

14.39 **Compare bumpers**:

a) $3.182(2.00)\sqrt{\dfrac{1}{2}+\dfrac{1}{2}} = 6.4$

b) The confidence interval formed using the Tukey 95% multiple comparison uses a 95% confidence level for the overall set of intervals, rather than a 95% confidence level for each separate interval.

c) Let $x_1 = 1$ for Bumper A and 0 otherwise, $x_2 = 1$ for Bumper B and 0 otherwise, and $x_1 = x_2 = 0$ for Bumper C.

d) 13 is the estimated mean damage cost for Bumper C, 11 is the difference between the estimated mean damage costs between Bumpers A and C, and 10 is the difference between the estimated mean damage costs between Bumpers B and C.

14.41 Compare segregation means:

a) From software: 15.1

b) All intervals contain 0; therefore, no pair is significantly different.

14.43 Comparing therapies for anorexia:

a) Software gives the following 95% confidence intervals:
control and cog: (-7.5, 0.6)
family and cog: (-8.8, 0.3)
family and control: (-12.4, -3.0)
The differences between the control and cognitive treatments and the family and cognitive treatments include 0; it is plausible that there are no population mean differences in these two cases. The other interval (i.e., family and control) does not include 0. We can infer that the population mean weight change is greater among those who receive family therapy than among those in the control group.

b) Software gives the following Tukey 95% multiple comparison confidence intervals:
Difference between control and cognitive: (-8.3, 1.4)
Difference between family and cognitive: (-9.8, 1.3)
Difference between family and control: (-13.3, -2.1)
The interpretations are the same as for the intervals in (a). The intervals are wider because the 95% confidence level is for the entire set of intervals rather than for each interval separately.

14.45 House location:

a) Let $x_1 = 1$ for NW and 0 otherwise. From software: price = 106836 + 26483 NW. The intercept, 106,836, is the estimated mean selling price when a house is not in the NW region, and 26,483 is the difference in the estimated mean selling price between a house in the NW region and a house not in the NW.

b) From software:

Predictor	Coef	SE Coef	T	P
NW	26483	12805	2.07	0.041

If the null hypothesis were true, the probability would be 0.04 of getting a test statistic at least as extreme as the value observed. We have considerable evidence that the population mean house selling price is significantly higher in the NW region than in the other regions.

c) From software:

Source	DF	SS	MS	F	P
NW	1	13149968133	13149968133	4.28	0.041
Error	98	3.01283E+11	3074311750		
Total	99	3.14433E+11			

As with the regression model, the P-value is 0.04; this is the same result.

d) The value of t in (b) is the square root of the value of F in (c).

14.47   Regress kids on gender and race:
a)   The coefficient of f (0.04) is the difference between the estimated means for men and women for each level of race. The fact that this estimate is close to 0 indicates that there is very little difference between the means for men and women, given race.
b)

	Female (1)	Male (0)
Black (1)	2.83	2.79
White (0)	2.46	2.42

c)   The data suggest that there is an effect only of race. If the null hypothesis were true, the probability would be close to 0 of getting a test statistic at least as extreme as the value observed. We have strong evidence that on the average blacks report a higher ideal number of children than whites do (by 0.37).

14.49   Regress TV watching on gender and religion:
a)   The mean estimate for men is 0.2 higher than the mean estimate for women at each fixed level of religion.
b)   The mean estimate for Protestants is 0.5 higher than the mean estimate for other religions at each fixed level of gender.
c) $\mu_y = \alpha + \beta_1 g + \beta_2 r_1 + \beta_3 r_2 + \beta_4 r_3$; for the response variable to be independent of religion, for each gender, β_2, β_3, and β_4 would need to equal 0.

14.51   Birth weight, age of mother, and smoking:
This suggests an interaction since smoking status has a different impact at different fixed levels of age. There is a bigger mean difference between smokers and non-smokers among older women than among younger women.

14.53   Salary and gender:
a)   The coefficient for gender, 0.8, indicates that at fixed levels of race, men have higher estimated mean salaries than women by 0.8 (thousands of dollars).
b)   (i) 31 (calculation: 30.2 + 0.8)   (ii) 30.8 (calculation: 30.2 + 0.6)

CHAPTER PROBLEMS: CONCEPTS AND APPLICATIONS

▭14.55 Number of friends and degree:
The short report will be different for each student, but can include and interpret the F test statistic of 0.81 and the P-value of 0.52.

14.57   A=B and B=C, but A ≠ C?:
There are many possible means that would meet these requirements. An example is means 10, 17, 24 for A, B, C.

14.59   Another Simpson paradox:
a)

	Female	Male
Humanities	56,000	55,000
Science	70,000	69,000

The overall mean for women is [25(56,000) + 5(70,000)]/30 = 58,333
The overall mean for men is [20(55,000) + 30(69,000)]/50 = 63,400
b)   A one-way comparison of mean income by gender would reveal that men have a higher mean income than women do. A two-way comparison of mean incomes by gender, however, would show that at fixed levels of university divisions, women have the higher mean.

14.61   ANOVA variability:
The best answer is (c).

14.63 Interaction:
The best answer is (c).

♦♦14.65 What causes large or small F?:
a) The four sample means would have to be identical.
b) There would have to be no variability (standard deviation of 0) within each sample. That is, all five individuals in each sample would have to have the same score.

♦♦14.67 Bonferroni multiple comparisons:
a) f: $1.88 \pm 2.46(0.7471)$, which is $(0.04, 3.7)$
   m: $1.96 \pm 2.46(0.7471)$, which is $(0.1, 3.8)$
b) $0.05/35 = 0.0014$

♦♦14.69 Regression or ANOVA?:
a) (i) In an ANOVA F test the number of bathrooms would be treated as a categorical variable. Three would not be considered "more" than one; it would simply be treated as a different category.
   (ii) In a regression t-test, the number of bathrooms would be treated as a quantitative variable and we would be assuming a linear trend.
b) The straight-line regression approach would allow us to know whether increasing numbers of bathrooms led to an increasing mean house selling price. The ANOVA would only let us know that the mean house price was different at each of the three categories. Moreover, we could not use the ANOVA as a prediction tool for other numbers of bathrooms (although even with regression, we must be careful when we interpolate or extrapolate).
c) Mean selling price $150,000 for 1 bathroom, $100,000 for 2 bathrooms, and $200,000 for 3 bathrooms. (There is not an increasing or decreasing overall trend.)

CHAPTER PROBLEMS: STUDENT ACTIVITIES

14.71 Student survey data:
The short reports will be different for each class.

Chapter 15
Nonparametric Statistics

SECTION 15.1: PRACTICING THE BASICS

15.1 **Tanning experiment:**

a)

Treatment	Ranks					
Lotion	(1,2)	(1,3)	(1,4)	(2,3)	(2,4)	(3,4)
Studio	(3,4)	(2,4)	(2,3)	(1,4)	(1,3)	(1,2)

b)

Lotion mean rank	1.5	2.0	2.5	2.5	3.0	3.5
Studio mean rank	3.5	3.0	2.5	2.5	2.0	1.5
Difference of mean ranks	-2.0	-1.0	0.0	0.0	1.0	2.0

c)

Difference between Mean Ranks	Probability
-2.0	1/6
-1.0	1/6
0.0	2/6
1.0	1/6
2.0	1/6

15.3 **Comparing clinical therapies:**

a) and b) together

Treatment	Ranks					
Therapy 1	(1,2,3)	(1,2,4)	(1,2,5)	(1,2,6)	(1,3,4)	(1,3,5)	(1,3,6)
Therapy 2	(4,5,6)	(3,5,6)	(3,4,6)	(3,4,5)	(2,5,6)	(2,4,6)	(2,4,5)
Therapy 1 mean rank	2.0	2.33	2.67	3.0	2.67	3.0	3.33
Therapy 2 mean rank 5.0	4.67	4.33	4.0	4.33	4.0	3.67	
Difference of mean ranks	-3.0	-2.33	-1.67	-1.0	-1.67	-1.0	-0.33

Treatment	Ranks					
Therapy 1	(1,4,5)	(1,4,6)	(1,5,6)	(2,3,4)	(2,3,5)	(2,3,6)	(2,4,5)
Therapy 2	(2,3,6)	(2,3,5)	(2,3,4)	(1,5,6)	(1,4,6)	(1,4,5)	(1,3,6)
Therapy 1 mean rank	3.33	3.67	4.0	3.0	3.33	3.67	3.67
Therapy 2 mean rank 3.67	3.33	3.0	4.0	3.67	3.33	3.33	
Difference of mean ranks	-0.33	0.33	1.0	-1.0	-0.33	0.33	0.33

Treatment	Ranks					
Therapy 1	(2,4,6)	(2,5,6)	(3,4,5)	(3,4,6)	(3,5,6)	(4,5,6)
Therapy 2	(1,3,5)	(1,3,4)	(1,2,6)	(1,2,5)	(1,2,4)	(1,2,3)
Therapy 1 mean rank	4.0	4.33	4.0	4.33	4.67	5.0
Therapy 2 mean rank 3.0	2.67	3.0	2.67	2.33	2.0	
Difference of mean ranks	1.0	1.67	1.0	1.67	2.33	3.0

c)

Difference between Mean Ranks	Probability
-3.00	1/20
-2.33	1/20
-1.67	2/20
-1.00	3/20
-0.33	3/20
0.33	3/20
1.00	3/20
1.67	2/20
2.33	1/20
3.00	1/20

d) The P-value is 4/20 = 0.20 (2/20 for each tail); if the treatments had identical effects, the probability would be 0.10 of getting a sample like we observed, or even more extreme, in either direction. It is plausible that the null hypothesis is correct and that the treatments do not lead to different results.

15.5 Estimating hypnosis effect:
a) The point estimate of -0.515 is an estimate of the difference between the population median ventilation for the control group and the population median ventilation for the treatment group.
b) The confidence interval of (-1.01, 0.02) estimates that the population median ventilation for the control group is between 1.01 below and 0.02 above the population median ventilation for the treatment group. Because 0 falls in the confidence interval, it is plausible that there is no difference between the population medians for the two groups.

15.7 Teenage anorexia:
a) The estimated difference between the population median weight change for the cognitive behavioral treatment group and the population median weight change for the control group is 3.05.
b) The confidence interval of (-0.6, 8.1) estimates that the population median weight change for the cognitive-behavioral group is between 0.6 below and 8.1 above the population median weight change for the treatment group. Because 0 falls in the confidence interval, it is plausible that there is no difference between the population medians for the two groups.
c) P-value = 0.11 for testing against the alternative hypothesis of different expected mean ranks. If the null hypothesis were true, the probability would be 0.11 of getting a test statistic at least as extreme as the value observed. It is plausible that the null hypothesis is correct and that the population distributions are identical.

SECTION 15.2: PRACTICING THE BASICS

15.9 What's the best way to learn French?:
a) Group 1 ranks: 2, 5, 6
   Group 2 ranks: 1, 3.5
   Group 3 ranks: 3.5, 7, 8
   The mean rank for Group 1 = (2+5+6)/3 = 4.33
b) P-value = 0.21; if the null hypothesis were true, the probability would be 0.21 of getting a test statistic at least as extreme as the value observed. It is plausible that the null hypothesis is correct and that the population median quiz score is the same for each group.

15.11 Cell phones and reaction times:
a) The observations are dependent samples. All students receive both treatments.
b) The sample proportion is 26/32 = 0.8125.
c) $se = \sqrt{(0.50)(0.50)/n} = \sqrt{(0.50)(0.50)/32} = 0.088$

   $z = (\hat{p} - 0.50)/se = (0.8125 - 0.50)/0.088 = 3.55$; P-value = 0.0002. If the null hypothesis were true, the probability would be 0.0002 of getting a test statistic at least as extreme as the value observed. We have strong evidence that the population proportion if drivers who have a faster reaction time when not using a cell phone is greater than 0.50.
d) The parametric method would be the matched-pairs *t*-test. The sign test uses merely the information about *which* response is higher and *how many*, not the quantitative information about *how much* higher. This is a disadvantage compared to the matched-pairs *t* test which analyzes the mean of the differences between the two responses.

15.13 Does exercise help blood pressure?:

H_0: $p = 0.50$. H_a: $p > 0.50$; all three subjects show a decrease. $P(3) = (0.50)^3 = 0.125$; if the null hypothesis were true, the probability would be 0.125 of getting a test statistic at least as extreme as the value observed. It is plausible that the null hypothesis is correct and that walking does not lower blood pressure (but we can't get a small P-value with such a small *n* for this test).

15.15 **More on cell phones:**

a) H_0: population median of difference scores is 0; H_a: population median of difference scores is > 0.

b)

Subject	1	2	3	4	Sample 5	6	7	8	Rank of absolute value
1	32	-32	32	32	32	-32	-32	-32	1
2	67	67	-67	67	67	-67	67	67	2
3	75	75	75	-75	75	75	-75	75	3
4	150	150	150	150	-150	150	150	-150	4

Sum of ranks for Positive differences	10	9	8	7	6	7	6	5

Subject	9	10	11	Sample 12	13	14	15	16	Rank of absolute value
1	32	32	32	-32	-32	-32	32	-32	1
2	-67	-67	67	-67	-67	67	-67	-67	2
3	-75	75	-75	-75	75	-75	-75	-75	3
4	150	-150	-150	150	-150	-150	-150	-150	4

Sum of ranks for Positive differences	5	4	3	4	3	2	1	0

c) P-value = 1/16 = 0.06; if the null hypothesis were true, the probability would be 0.06 of getting a test statistic at least as extreme as the value observed. There is some, but not strong, evidence that cell phones tend to impair reaction times.

CHAPTER PROBLEMS: PRACTICING THE BASICS

15.17 **Car bumper damage:**

a) Bumper A: ranks are 4, 6, 5; mean is 5.
Bumper B: ranks are 1, 2, 3; mean is 2.

b)

Treatment	Ranks
Bumper A	(1,2,3) (1,2,4) (1,2,5) (1,2,6) (1,3,4) (1,3,5) (1,3,6) (1,4,5) (1,4,6) (1,5,6)
Bumper B	(4,5,6) (3,5,6) (3,4,6) (3,4,5) (2,5,6) (2,4,6) (2,4,5) (2,3,6) (2,3,5) (2,3,4)

Treatment	Ranks
Bumper A	(2,3,4) (2,3,5) (2,3,6) (2,4,5) (2,4,6) (2,5,6) (3,4,5) (3,4,6) (3,5,6) (4,5,6)
Bumper B	(1,5,6) (1,4,6) (1,4,5) (1,3,6) (1,3,5) (1,3,4) (1,2,6) (1,2,5) (1,2,4) (1,2,3)

c) There are only two ways in which the ranks are as extreme as in this sample: Bumper A with 1,2,3 and B with 4,5,6, or Bumper A with 4,5,6 and B with 1,2,3.

d) The P-value is 0.10 because out of 20 possibilities, only two are this extreme. 2/20 = 0.10.

15.19 **Telephone holding times:**

a)

Group	Ranks	Mean Rank
Muzak	1,2,4,5,3	3.0
Classical	9,8,7,10,6	8.0

b) There are only two cases this extreme, that in which Muzak has ranks 1-5 as it does here, and that in which Muzak has ranks 6-10. Thus, the P-value is the probability that one of these two cases would occur out of the 252 possible allocations of rankings. If the treatments had identical effects, the probability would be 0.008 of getting a sample like we observed or even more extreme. This is below a typical significance level such as 0.05; therefore, we can reject the null hypothesis.

15.21 **Comparing tans:**

a) Kruskal-Wallis test

b) There are several possible examples, but all would have one group with ranks 1-3, one with ranks 4-6 and one with ranks 7-9.

15.23 **Internet vs cell phones:**
    a)   (i) H_0: Population proportion $p = 0.50$ who use cell phones more than the Internet; H_a: $p \neq 0.50$

         (ii) $se = \sqrt{(0.50)(0.50)/n} = \sqrt{(0.50)(0.50)/39} = 0.080$

           $z = (\hat{p} - 0.50)/se = (0.897 - 0.50)/0.080 = 4.96$

         (iii) P-value = 0.000; if the null hypothesis were true, the probability would be near 0 of getting a test statistic at least as extreme as the value observed. We have extremely strong evidence that a majority of countries have more cell phone use than Internet use.

    b) This would not be relevant if the data file were comprised only of countries of interest to us. We would know the population parameters so inference would not be relevant.

15.25 **GPAs:**
    a) We could use the sign-test for matched pairs or the Wilcoxon signed-ranks test.
    b) One reason for using a nonparametric method is if we suspected that the population distributions were not normal, for example, possibly highly skewed, because we have a one-sided alternative, and parametric methods are not then robust.

    c) From software (used CGPA-HSGPA):

```
Test of median = 0.000000 versus median < 0.000000
                     N
                    for   Wilcoxon           Estimated
         N     Test   Statistic    P          Median
C20     59      55       268.5   0.000       -0.1850
```

If the null hypothesis were true, the probability would be near 0 of getting a test statistic at least as extreme as the value observed. We have very strong evidence that population medain high school GPA is higher than population median college GPA.

15.27 **Wilcoxon signed-rank test about the diet:**
    a) H_0: population median of difference scores is 0; H_a: population median of difference scores > 0.

Subject	**Sample**								Rank of absolute value
	1	**2**	**3**	**4**	**5**	**6**	**7**	**8**	
1	120	120	-120	120	-120	120	-120	-120	3
2	15	15	15	-15	15	-15	-15	-15	2
3	2	-2	2	2	-2	-2	2	-2	1

Sum of ranks for positive differences   6   5   3   4   2   3   1   0

The rank sum is 5 1/8 of the time, and is more extreme (i.e., 6) 1/8 of the time. Thus, the P-value = 2/8 = 0.25. If the null hypothesis were true, the probability would be 0.25 of getting a test statistic at least as extreme as the value observed. It is plausible that the null hypothesis is correct and that the population median of difference scores is not positive.

    b) The results are identical to those in Example 8. Outliers do not have an effect on this nonparametric statistical method.

CHAPTER PROBLEMS: CONCEPTS AND INVESTIGATIONS

15.29 **Why nonparametrics?:**
There are many possible situations. One example is a situation in which the population distribution is likely to be highly skewed and the researcher wants to use a one-sided test.

15.31 **Complete the analogy:**
Kruskal-Wallis

15.33 **True or false?:**
False

♦♦**15.35 Mann-Whitney statistic:**
    a)   The proportions are calculated by pairing up the subjects in every possible way, and then counting the number of pairs for which the tanning studio gave a better tan. For the first set of ranks in the chart (lotion: 1,2,3 and studio: 4,5), the possible pairs are as follows with lotion first: (1,4) (1,5) (2,4) (2,5) (3,4) (3,5). In none of these pairs did the studio have the higher rank; therefore, the proportion is 0/6.

    b)

Proportion	Probability
0/6	1/10
1/6	1/10
2/6	2/10
3/6	2/10
4/6	2/10
5/6	1/10
6/6	1/10

    c)   The P-value is 2/10. The probability of an observed sample proportion of 5/6 or more extreme (i.e., 6/6) is 2/10 = 0.20.

♦♦**15.37 Nonparametric regression:**
The nonparametric estimate of the slope is not strongly affected by a regression outlier because we are taking the median of all slopes. The median is not susceptible to outliers. The ordinary slope, on the other hand, takes the magnitude of all observations into account.

REVIEW PROBLEMS: PRACTICING THE BASICS

R4.1   **Gender and opinion about abortion:**
a)

	Unrestricted abortion should be legal?	
Gender	Yes	No
Male	0.40	0.60
Female	0.40	0.60

b)   Independent since the opinion of the respondent is the same regardless of the respondent's gender.

R4.3   **Murders and gender:**
a)   Output from Minitab follows:
```
Expected counts are printed below observed counts
Chi-Square contributions are printed below expected counts

        Female    Male   Total
  1        185     524     709
        196.69  512.31
         0.695   0.267

  2       1793    4628    6421
        1781.31 4639.69
         0.077   0.029

Total    1978    5152    7130
Chi-Sq = 1.068, DF = 1, P-Value = 0.301
```
The test statistic is $\chi^2 = 1.07$ with a P-value of 0.30.

b)   The difference in the proportions of male versus female offenders when the victim was female is (1793/6421)-(185/709)=0.28-0.26=0.02. The difference in the proportions of male versus female offenders when the victim was male is (4628/6421)-(524/709)=0.72-0.74= -0.02.

R4.5   **Ma and Pa Education:**
The entry for row i, column j gives the correlation between the variable in row i and the variable in column j. The correlation tells us the strength of the linear association between two variables. Note that when i=j, the correlation is 1 since every variable is perfectly correlated with itself. Mother's education and father's education have the strongest linear association with a correlation of 0.65. All of the linear associations are positive (as one variable increases, so does the other tend to increase) since all of the correlations are positive.

R4.7   **Fertility and contraception:**
a) $\hat{y} = 6.663 - 0.06484x$

(i) $\hat{y} = 6.663 - 0.0648(0) = 6.66$

(ii) $\hat{y} = 6.663 - 0.0648(100) = 0.18$

We can obtain the difference between these by using the slope. The difference is the change in x (100-0) times the estimated slope (-0.0648).

b) $\hat{y} = 6.663 - 0.0648(51) = 3.4$; the residual is $y - \hat{y} = 1.3 - 3.4 = -2.1$

c)   The standardized residual of -2.97 indicates that the observation for Belgium was 2.97 standard errors below the predicted value from the regression line.

R4.9 **Predicting body fat:**
a) r^2 is 0.89. This indicates a strong association.
b) The P-value of 0.000 indicates that it if H_0 were true that the population slope $\beta = 0$, it would be extremely unusual to get a sample slope as far from 0 as $b = 1.338$.
c) The 95% confidence interval is in agreement because 0 is not included in the interval and is therefore not a plausible value for the slope.
d) The residual standard deviation is the square root of the residual *MS*. The square root of 2.7 is 1.64. The residual standard deviation describes the typical size of the residuals and estimates the standard deviation of y at a fixed value of x.

R4.11 **Using the Internet?:**
a) No, the slopes are in different units.
b) Yes, the slopes measure the change in the percentage of those using either cell phones or the Internet for a 1 unit ($1000) change in GDP. The first slope, 2.62, predicts that for every $1000 increase in GDP, the percentage using cell phones will increase by 2.62 and the second tells us that for every $1000 increase in GDP, the percentage using the Internet will increase by 1.55.

R4.13 **Growth of Wikipedia:**
a) 10,000 represents the predicted number of English-language articles in Wikipedia as of January 1, 2003. 2.1 represents the estimated multiplicative effect on the mean of y for each one year change in x.
b) (i) $\hat{y} = 100,000(2.1)^5 = 4,084,101$

  (ii) $\hat{y} = 100,000(2.1)^{10} = 166,798,810$

  2013 is too far into the future from the observed data on which the model was built to trust the prediction, 2008 may be as well. Since posting articles on the Internet is a relatively new practice, it is possible that the number of articles posted in a given time period will level off, thereby changing the relationship between x and y. Caution should be used in making predictions into the future based on this model.

R4.15 **Baseball offensive production:**
a) Since the coefficient of HR has the greatest magnitude (1.48), it has the largest effect on $\hat{y}$ for a one-unit increase.
b) As the number of stolen bases increases, the predicted number of runs scored will increase, since the coefficient for SB is positive. The coefficient for CS is negative, so that as the number caught stealing increases, the predicted number of runs scored will decrease.
c) $\hat{y} = 100 + 0.59(600) + 0.71(100) + 0.91(10) + 1.48(200) + 0.30(300) + 0.27(40) - 0.14(4000) - 0.20(20) = 367$.

R4.17 **Predicting pollution:**
a) R=0.787 is the correlation between the observed CO2 values and those predicted using GDP and life expectancy as predictors.
b) No. The multiple correlation increased only very slightly. Adding life expectancy did very little, if anything, to increase the predictive power of the model.
c) It is possible that life expectancy and GDP are strongly associated, so that once one of these variables is in the model, adding the other does not do much to increase the predictive power of the model. Thus, just because the multiple correlation did not increase by much, this does not imply that the additional variable is weakly correlated with the response variable.
d) The model predicts the mean CO2 values for each set of GDP and life expectancy values within range. A large positive standardized residual means that the observed GDP value is larger than the value predicted by the multiple regression model at the given levels of GDP and life expectancy. Australia's standardized residual is 2.84; thus, Australia's observed CO2 level was 2.84 standard deviations above the value predicted by the multiple regression model using GDP and life expectancy as predictors.

R4.19   **Interaction between SES and age in quality of health:**

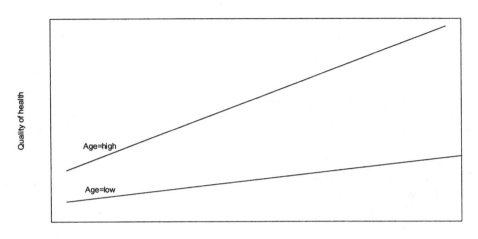

R4.21   **Political ideology and party affiliation:**
  a)   The response variable is political ideology score. The factor is the respondent's political party.
  b) H_0: $\mu_1 = \mu_2 = \mu_3$; H_a: at least two of the population means are unequal; μ_1 denotes the population mean political ideology score for Democrats; μ_2 denotes the population mean political ideology score for Independents and μ_3 indicates the population mean political ideology score for Republicans.
  c)   Since the F statistic has a corresponding P-value of approximately 0, there is sufficient evidence to reject the null hypothesis. The probability of getting an F statistic at least as large as that observed is essentially 0 under the null hypothesis. We can conclude that the population mean political ideology scores differ for at least two of the political parties.
  d)   Independent random samples from normal population distributions with equal standard deviations. These seem plausible

R4.23   **Income and education:**
  a)   (i) $\hat{y} = 20 + 23(0) + 17(0) = 20$; (ii) $\hat{y} = 20 + 23(1) + 17(0) = 43$.
  b)   The estimated population mean income in 2005 for college graduates is $23,000 higher than for high school graduates when gender is held constant.
  c)   (ii), because there are two categorical predictors of income.

REVIEW PROBLEMS: CONCEPTS AND INVESTIGATIONS

⌨R4.25 Racial prejudice:
  1)   Assumptions: random sampling was used and the sample sizes are large enough so that the expected cell counts are at least 5.
  2) H_0: religious preference and whether or not one favors laws against interracial marriage are independent
        H_a: religious preference and whether or not one favors laws against interracial marriage are dependent
  3)   From Minitab: $\chi^2 = 28.2$.
  4)   P-value=0.000.
  5)   If the variables are independent, the probability of obtaining a test statistic at least as large as that observed is essentially 0. Thus, we reject the null hypothesis and conclude that whether or not one favors laws against interracial marriage depends on their religious preference.
        Also, can do residual analysis and compare groups using difference of proportions.

⌨R4.27 Analyze your data:
        Answers will vary.

R4.29 **Predicting college success:**
This could summarize the result of a regression analysis using college GPA as the response and the various factors considered by the admissions officers as the explanatory variables. The 30% refers to R^2 and gives the proportional reduction in error using $\hat{y}$ to predict y rather than using $\bar{y}$ to predict y.

R4.31 **Why ANOVA?:**
One-way ANOVA is a method used to compare the population means of several groups. The null hypothesis is that all of the groups come from populations with equal means and the alternative is that at least two of these population means differ. If the null hypothesis is rejected, we can compute confidence intervals for pairs of means to determine which population means are different as well as how different they are.

R4.33 **Violating regression assumptions:**
a) Exponential regression model
b) Logistic regression model